Islands of Resistance

Islands of Resistance

Pirate Radio in Canada

edited by
Andrea Langlois, Ron Sakolsky,
& Marian van der Zon

NEW STAR BOOKS • VANCOUVER • 2010

Islands of Resistance

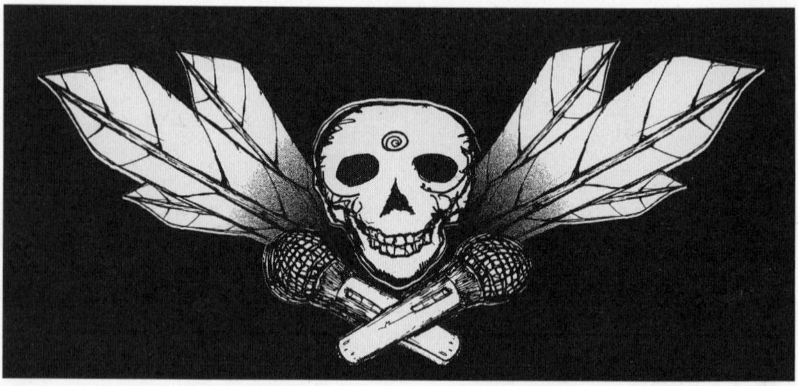

Contents

1. **Setting Sail: Navigating Pirate Radio Waves in Canada**
   Ron Sakolsky, Marian van der Zon and Andrea Langlois — 3

2. **Latitudes of Rebellion: Free Radio in an International Context**
   Stephen Dunifer — 23

3. **Resistance to Regulation Among Early Canadian Broadcasters and Listeners**
   Anne F. MacLennan — 35

4. **Freedom Soundz: A Programmers Journey Beyond Licensed Community Radio**
   Sheila Nopper — 51

5. **Secwepemc Radio: Reclamation of Our Common Property**
   Neskie Manuel — 71

6. **Awakening the 'Voice of the Forest' Radio Barriere Lake**
   Charles Mostoller — 75

7. **Squatting the Airwaves: Pirate Radios Anarchy in Action**
   Ron Sakolsky — 89

8. **Amplifying Resistance: Pirate Radio a Protest Tactic**
   Andrea Langlois and Gretchen King — 101

9. **The Care and Feeding of Temporary Autonomous Radio**
   Marian van der Zon   117

10. **The Voyage of a Gender Pirate and Her Toolbox**
    Bobbi Kozinuk   133

11. **Pirate Radio and Maneuver: Radical Artistic Practices in Quebec**
    André Éric Létourneau (translation by Clara Gabriel)   145

12. **Touch That Dial: Creating Radio Transcending the Regulatory Body (1990)**
    Christof Migone   161

13. **The Art of Unstable Radio**
    Anna Friz   167

14. **Repurposed and Reassembled: Waking Up the Radio**
    Kristen Roos   183

15. **Radio Ballroom Halifax**
    Stephen Kelly and Eleanor King (with Marian van der Zon)   195

16. **The Power of Small: Integrating Low-Power Radio and Sound Art**
    Kathy Kennedy   201

17. **Voices in a Public Place: A Docudrama in Seven Acts on/for Micro-radio in Canada**
    Roger Farr   213

Rip-Roaring Radical Radio Resources   229
Bibliography   231
The Contributors   238

List of Illustrations

Cover: Maurice Spira
Table of Contents: Charlie the Headcase (illustration)
Chapter 1: Moss Dance (illustration) www.rainbowraven.ca
Chapter 2: Stephen Dunifer (photograph)
Chapter 3: Duncan Macphail, (photograph) Source: in Duffy, Dennis. Imagine Please: Early Radio Broadcasting in British Columbia. (Victoria: Provincial Archives of British Columbia, 1983. 64.)
Chapter 4: Matta (illustration)
Chapter 5: Gord Hill (illustration)
Chapter 6: Image 1: Charles Mostoller (photograph); Image 2: Charles Mostoller (photograph)
Chapter 7: Tomás Hayek (collage)
Chapter 8: Gretchen King (photograph)
Chapter 9: Image 1: Sandra Morlacci (illustration); Image 2: Gary Eugene (photograph)
Chapter 10: "Wild and Infinite Flight" by Anais LaRue (illustration)
Chapter 11: Image 1: Photomontage by Phillipe, still photography by Jeff, courtesy of Diffusion système minuit du Québec archive center, Laval, Quebec (photomontage); Image 2: Photo taken by Pouf's brother (photograph); Image 3: Abribec, suppôt de la novelle humanité fiscale (http://www.iso1000000000.ch/abribec) (illustration)

Chapter 12: "Absolute Final Intimacy of an Ear" by Chris McClaren (collage)

Chapter 13: Image 1: Alexis O'Hara (photograph); Image 2: Claire Pfeiffer (illustration)

Chapter 14: Kristen Roos (illustration)

Chapter 15: Andrea Lalonde (photograph)

Chapter 16: Image 1: Destanne Lundquist (collage); Image 2: Kathryn Walker (photograph); Image 3: Kathy Kennedy (illustration of score)

Chapter 17: Jesse Gentes (collage)

RRR Resources: Beehive Design Collective, anti-copyright, www.beehivecollective.org (illustration); Vrindavanesvari Conroy (illustration)

Acknowledgements

The editors wish to acknowledge the Black Ink BC Anarchist Writers' Group and the Inner Island retreat where the idea for this book was born. We extend our gratitude to New Star Books: Rolf Maurer, Stefania Alexandru, Jamie Nadel and Michael Barnholden, for their support of this project. A huge thanks to all of our intrepid authors and incredible artists! We thank the ever fastidious Clara Gabriel for her work in translation. We acknowledge our pirate comrades Dennis Burton (Radio Radio), Peter Cloud Panjoy (Free Radio Hornby), Rob Schmidt (Free Radio Sundridge), Sean Coyote (Free Radio Cortes), Susan Kennard (Radio 90) for their early participation in this project. Finally, for their researching and networking tips, a tip of the hat to: Afua Cooper, Bob Sarti, Chantal Gutstein, Frederic Dubois, Gayle Young, Jane Parkinson, Jo Mrozewski, John Harris Stevenson, Leslie Regan Shade, Lorna Roth, Marc Raboy, Mary Vipond, Michel Sénécal, Naava Smolash, Pam Bullick, Sarita Ahooja and Steve Anderson. Ron thanks Sheila Nopper for her indomitable pirate spirit. Marian thanks Gary Eugene for his patience, direct action and piratical support. She extends this thanks to her family and friends, particularly to Asta, Alex and Linda. Andrea thanks her friends and family (especially Mom, Dad, Rick, Mélanie and Clara), and Ken Tupper for jumping aboard her pirate ship with his dictionary and spirited love of ideas.

MOSS DANCE

CHAPTER 1

Setting Sail

Navigating Pirate Radio Waves in Canada

Ron Sakolsky, Marian van der Zon
and Andrea Langlois

JOIN US AS WE SET SAIL FOR THOSE ISLANDS OF RESISTance known as pirate radio. Bypassing the treacherous waters of licensing and the doldrums of institutionalization, we have eschewed the fixed maps of entrenched power in favour of a cartography of autonomy. As navigators along these diverse routes, we have been guided by an illuminated chart composed of pirate ports of call burning as brightly as the star constellations relied upon by all mariners at sea.

We begin our voyage in concrete terms — what is "pirate radio"? The use of the term pirate radio is a controversial one. It has been burdened with the negative connotations of theft and mayhem, and exoticized with romantic swashbuckling imagery and Hollywood production values. Perhaps the term "free radio" would carry less baggage, but we chose pirate radio because it is more immediately understood by North Americans as referring to an unlicensed form of radio broadcasting that relies on the airwaves for transmission, rather than the internet-based mechanisms of podcasting or web radio. As Anne MacLennan points out (Chapter 3), the word pirate had a pejorative ring to it even in the 1920s and 1930s, being associated with predatory US broadcasters overriding the signals of Canadian-based stations. While this particular connotation might still exist in some quarters, for us it is the "no quarters" transgressive quality of the word pirate that we embrace as inspiring to both the radical imagination and the practice of direct action.

If pirate is a word that has been used disparagingly by the radio industry and its counterparts at the Canadian Radio-television Telecommunications Commission (CRTC), then we hereby reclaim it as a badge of honour. We prize the fact that it evokes an outsider status[1] in relation to the dominant cultural assumptions and practices of the radio industry and the government bodies that are theoretically supposed to regulate the airwaves in the "public interest." As to those regulatory agencies, in practice they have been captured by the very same vested interests within the radio industry that they are mandated to oversee. The legal climate in which they operate facilitates the corporate theft of the airwaves and squelches autonomous alternatives with a bureaucratic arsenal of administrative rules and sanctions.

In essence, what all radio pirates have in common is a refusal to obey these legal edicts, whether out of a sense of political injustice, a defiant libertarian ethos, a desire for self-realization or for purposes of artistic expression. In this regard, what links the radio pirates in this book is that their projects can all be placed on a spectrum of illegality. On such a spectrum, illegality is viewed in a positive light rather than vilified and dismissed. It is the transgressive nature of radio piracy — or "electromagnetic deviance" as André Éric Létourneau (Chapter 11) describes it — that concerns us here. Transgression can take many forms, from intentional interference with licensed broadcasts to floating islands of sporadic insurrection and temporary autonomous zones, as well as more permanently-situated islands of resistance that are rooted in geographical, ethnic, gendered or culturally-based forms of community.

It is the interrelations and interactions among these transgressive pieces of the pirate radio puzzle that allow us to understand the larger picture. The nomadic radio pirate strategically broadcasting the location and movements of the police to global justice activists during the heat of confrontation in the streets engages in direct action by occupying the airwaves. So too is the stationary pirate radio broadcaster reporting on these events and their context to her community, as is the radio artist whose unlicensed broadcasts involve an aesthetic subversion that reimagines the theory and practice of radio in public spaces. While someone could argue that one particular tactical use of radio is more or less transgressive than another based on the type of illegality involved, we prefer to focus upon what might be called the "transgressive trace" that animates them all.

An interesting rubric that can be used to conceptualize the emancipatory potential of such radio transgressions is what Stevphen Shukaitis calls a "minor cultural politics." Here minor is not meant to diminish the importance of such a politics, but rather to situate it as a form of self-organization that may not be as grandiose as the all or nothing quality that revolutionary rhetoric allows, but which has serious implications in liberating radical possibilities. "It is this form of politics based not upon projecting an already agreed upon political solution or calling upon an existing social subject (the people, the workers), but rather developing a mode of collective, continual and intensive engagement with the social world that embodies the politics of minor composition."[2] Shukaitis does not reference pirate radio in particular as a form of minor cultural politics, but he does point to the oppositional aspects of punk culture (particularly invoking the subversive irony embedded in the name of the band Minor Threat).

Over the years many other theoretical approaches have been used to conceptualize transgressive radio, several of which are personified within Roger Farr's docudrama "Voices in a Public Space" (Chapter 17). One of these theoretical wellsprings we wish to highlight here is the work of Felix Guattari and Franco "Bifo" Berardi, who were co-conspirators at the legendary Radio Alice in Italy during the heady days of the Autonomia movement of the seventies. It was their experiences at Radio Alice that inspired them to call for the destruction of the hierarchical and centralized mass media model in favour of a de-territorialized proliferation of diverse and criss-crossing "enunciative" possibilities that they termed "popular free radio." To accomplish this emancipatory vision, Guattari imagined a "post-mediatic" world based upon "molecular self-organization" that anticipated the most libertarian aspects of the internet and cyberculture.[3] Instead of the system of "semiocapitalism," Berardi envisioned "the creation of a new public space, autonomous from both state monopoly and private economic domination."[4] In keeping with his autonomist vision of telecommunications, Guattari chose the inspirational title "Millions and Millions of Potential Alices" for the preface to the French edition of the pamphlet *Alice is the Devil, Free Radio Alice*.

Yet Radio Alice's idea of free radio was vastly different from both that of earlier offshore pirate stations such as Radio Caroline in the

UK, which was a commercial broadcaster, and the kind of "alternative radio" that in subsequent years has characterized licensed campus/community radio stations in Canada. Extrapolating further on these differences in his discussion of Guattari's conception of Radio Alice, Michael Goddard noted:

> This is miles away both from ideas of local or community radio in which groups should have the possibility on radio to represent their particular interests and from conventional ideas of political radio in which radio should be used as a megaphone for mobilizing the masses ... What this type of radio achieved most of all was the short-circuiting of representation in both the aesthetic sense of representing the social realities they dealt with and in the political sense of the delegate or authorized spokesperson in favor of generating a space of direct communications.[5]

In this more expansive sense, Radio Alice actively challenged listener passivity and encouraged its audience members to become engaged in direct speech on the airwaves through relaying live call-ins aimed at unleashing strategic reports from the barricades, along with the unfiltered rage of protesters in the streets and the poetic laughter of the insurgent imagination in flight. And, for the playful protagonists and provocateurs of Radio Alice, if this could only be done illegally, then so much more the merrier. As it turned out in the end, the more sombre response of the Italian state was arresting and imprisoning Radio Alice's operators on charges of sedition, and shutting it down.

What then would a spectrum of illegality look like in relation to the stories of radio piracy that you will encounter in the following chapters? While we do not want to fetishize the degree of illegality as proof of radicalism, one way of envisioning this spectrum is in relation to the type of legal transgression involved. Not all forms of pirate radio involve the same perceived or actual risks. Within the wide array of programming that encompasses pirate initiatives from "permanent" radio stations to temporary broadcasters, it is a complicated task to decide where each pirate project fits on a spectrum of illegality. For example, the signal of a community-based pirate station on an indigenous reserve may be unlicensed and unwelcome in a nearby settler community, but it has sovereignty protections not available off-reserve. Similarly, a temporary broadcaster who is immersed in the tactical uses of pirate radio in protest situations seemingly faces a bigger risk of getting busted than a temporary broadcaster whose challenge to corporate and governmental stations might revolve around playing dance music not available over the licensed airwaves. Unless, of course, in the case of the latter, there is a complaint from the media moguls whose market share is threatened.

According to the CRTC, all unlicensed radio is illegal. Technically, this includes micro-radio, a form of narrowcasting, usually understood as being less than five watts, as well as low-power transmissions, which are generally less than 100 watts. How far a signal could be broadcast at two watts or 100 watts would depend on geography and antenna placement.[6] In the micro format, which is often used by radio artists, the risk of operating in Canada is typically interpreted as being very small. Beyond micro-radio, Canadian pirate radio in general occupies a niche that is unlicensed, and for the most part it flies under the radar; unless it is discovered accidentally by the authorities or there is interference with existing licensed broadcasters and a complaint is lodged. In this case, the pirate broadcaster's disruption of the established pecking order may be met with repression because legally sanctioned high-watt stations are automatically given a privileged position by the CRTC.

Though diverse in their scope, all of the radio pirates in this book constitute a living assemblage of the frequencies of resistance. When viewed in this light, they can be understood as collectively having the wide-ranging potential to inspire a future movement of radio piracy in Canada. If the links we make here help to kick off increased resistance to legal constraints and enhanced networking among radio resisters, then we will rejoice at the continuing power of both radio and the written word to subvert the established order.

While Canadian pirate radio initiatives may not, at present, constitute the kind of movement that has been in evidence in recent years among radio pirates in the United States, in some ways the role of Bobbi Kozinuk, one of the writers in this volume (Chapter 10), might be seen as analogous to that of American pirate radio activist and founder of Free Radio Berkeley, Stephen Dunifer, who also has authored a chapter in this book (Chapter 2). Like Dunifer, Kozinuk has facilitated numerous workshops on building micropower transmitters. These workshops have in turn been an impetus for the spread of pirate radio throughout Canada. However, though both have been influenced by the pioneering micropower radio work of Tetsuo Kogawa in Japan, because Kozinuk is involved to a larger extent than Dunifer in the radio art community, her transmitter building workshops have resulted in a greater proportion of Canadian pirates being sound artists than has been the case in the US. On the other hand, Dunifer's international impact among radicalized grassroots radio pirates is more widespread than that of Kozinuk. Yet, Kozinuk is not apolitical; she has used pirate radio as a tool in protest situations in Canada and continues to use pirate radio as a means to challenge social norms.

Moreover, Canadian radio pirates do not seem to have any interest in promoting the kind of nationalist agenda that is sold to the listeners of the Canadian Broadcasting Corporation (CBC) as "Canadian identity." They also typically resist the free market seductions of the corporate sector, which has as its main goal the selling of audiences (potential consumers) to advertisers. Most Canadian radio pirates typically enjoy being independent of both, and free to fly the Jolly Roger rather than the officially sanctioned flag of the nation state or the branded banner of the multinational corporation. However, not all radio pirates would consider themselves to be radical resisters or movement activists. Radio pirates take over the airwaves illegally for various reasons. Some do so to maintain language and culture, or as a statement of indigenous sovereignty. Others wish to create community or to protest domination. In some cases, the objective is to give direct voice to the voiceless by strengthening the self-defined identity of a singularly-minded group or of a group that is more inclusive of the complexity of different geographical, ethnic and gendered realities. For some, the primary motivation is to create art or to devise a space of self-representation in relation to music, politics, the spoken word, sound art or radio drama. For others, their dedication to autonomous radio goes beyond content and becomes a participatory experiment in lateral organization and facilitates access to the skills and tools needed for cultural production. Finally, some go on-air as pirates to exercise the individual freedom to use radio in experimental and unconventional ways, such as the method of repurposing explored by Kristen Roos (Chapter 14). In all cases, as pirates they exhibit the underlying philosophy that the airwaves should be freely available.

The context of who is a pirate radio practitioner has changed over time. Anne MacLennan (Chapter 3) documents how, in the 1920s and 1930s, competing border radio stations licensed in Mexico and the US were considered to be pirates by the CRTC because their signals overpowered those of Canadian stations. Today the definition of piracy is not so much about interfering with licensed broadcasts as it is about engaging in any unlicensed programming activity, though intentional interference still occurs on occasion by politicized pirate broadcasters as is illustrated by André Éric Létourneau (Chapter 11). In the early days of Canadian radio, regulation was sparse, and many pirates who took to the airwaves could slip between the loopholes of Canadian law. They were even joined in their rebellion against licensing by many audience members who refused to buy the required government-issued licences in order to listen to the Canadian broadcasting service.[7]

Indigenous Roots

Although most examples of pirate radio in this book are contained within the settler-defined borders of what is called Canada, we feel that it is important to indicate our discomfort with this demarcation. Accordingly, we acknowledge the many First Nations of Turtle Island and the fact that the reserve-based radio broadcasters written about here occupy or have occupied the airwaves that are part of colonized and/or unceded indigenous territory. In effect, these broadcasts serve to question and challenge the Canadian government's control of the airwaves. As Neskie Manuel (Chapter 5) explains, not applying for a CRTC licence to broadcast is an acknowledgement that the Secwepemc people "did not give up our right to make use of the electromagnetic spectrum to carry on our traditions, language and culture." Similar claims of sovereignty are expressed by the Algonquin broadcasters featured in the Charles Mostoller essay on Radio Barriere Lake (Chapter 6).

In this regard, it should be pointed out that Canadian radio pirates as a whole can trace their lineage back to the direct actions of indigenous peoples in asserting their sovereignty, although not all indigenous peoples who engage in unlicensed radio broadcasts would choose to call themselves pirates. Even in terms of the history of the earliest licensed community radio stations, indigenous radio played a seminal role. It was some of the staff involved in an unlicensed 1969 low-power mobile station called Radio Kenonashwin on Ojibway territory, Longlac, Ontario, who later offered crucial advice and support in the setting up of Vancouver Co-op Radio (CFRO) in the early 1970s.[8]

One example of how the sovereignty struggles of indigenous peoples have informed the historical development of pirate radio in a Canadian context is Akwesasne Freedom Radio, which was originally based at the Mohawk reserve on the St. Lawrence River (encompassing Ontario, Québec and New York State). Because the station crossed the US/Canadian border, it was able to resist being licensed by either the Canadian CRTC or the Federal Communications Commission (FCC) in the United States. Instead, it sought its approval to broadcast through a proclamation by the Akwesasne Mohawk Nation. As Charles Fairchild has pointed out, "This arrangement was acknowledged, but not controlled by the CRTC, while the FCC refused to even recognize the station itself."[9] However, when Akwesasne Freedom Radio went off the air, the Canadian federal government required its successor, CKON, to apply for a licence to operate in 1982. The traditional Mohawk council of chiefs then began negotiating with the CRTC for regulatory rights. Eventually, according to Michael Keith, the CRTC avoided a confrontation by "recognizing the Mohawk's ability to regulate the station's operation."[10] Legally, then, we are left with a confusing cross-border situation in which Keith says the CRTC "recognized" both the station and the regulatory power of the Mohawks, while Fairchild says the FCC "refused" to recognize the station at all. In any case, the CRTC was prevented from shutting down the station largely as a result of the kind of direct challenge to colonial authority by Mohawk defenders that would later be evidenced during the Oka conflict and the subsequent Kanesatake insurgency.

Positioning Pirate Radio in the Mediascape

While the indigenous concept of sovereignty is a very different construct from what in European terms is called the "commons," both challenge control of the airwaves by the corporate state. In the latter case, the airwaves are seen as an open public space where everyone should have equal access to the means of having their voice heard. In Canada, most media outlets — from public radio to newspapers — are woefully inadequate in terms of encompassing, much less directly representing, the diversity of voices and identities that exist throughout the country. Instead, the mediascape is saturated with news, music and culture tailored to market demographics and programming that embodies and perpetuates the omnipresent representations of dominant ideologies. When using the theoretical construct of the commons, which sees media as a social resource to be shared, the questions of who is being represented, and how, become paramount. Beyond notions of representative democracy, one can alternatively conceptualize underground organizational forms like pirate radio as being part of an "infrapolitical undercommons" that is often temporary and/or invisible by design in order to escape the gaze of power and prevent enclosure.[11]

As a radical media strategy, occupying the airwaves can be viewed as a way to break down the hierarchies of access to meaning-making that are characteristic of licensed radio, thereby allowing grassroots individuals and marginalized groups not only a voice, but also the ability to define their own realities. This constructive ability is associated with what Pierre Bourdieu refers to as "symbolic power."[12] As Nick Couldry points out, symbolic power is not evenly distributed in our society. This, he says, "has been an everyday fact of life in most societies, but it takes a particular form in contemporary mediated societies, where symbolic power is concentrated, particularly, although not of course exclusively, in media institutions, so that the uneven distribution of symbolic resources results in the overwhelming reality of media power."[13]

It is not simply the content and symbols presented in the mediasphere in and of themselves that indicate a concentration of power, but also the way that communication technologies have become centralized, marketed and developed in order to concentrate media power in the hands of a few. While accessible two-way communication — a form of communicating that breaks down the producer/audience dichotomy — has only recently been widely popularized through new technologies such as the internet, historically, as we have already seen, such an anti-authoritarian approach had been prefigured by pirate radio practitioners at Radio Alice. Even so, over thirty years later, most North Americans still have trouble conceptualizing the possibility of using the airwaves to engage in passionate two-way communication. For example, the one-way kind of call-in show format — now a fixture of mainstream radio — is a far cry from the freewheeling form that the call-in took at Radio Alice. Instead of the dual flow of Radio Alice, the type of call-in program with which most radio listeners are familiar is tightly controlled by host, producer and ultimately station management. Transgressing the boundaries of who has the ability to be heard by, and communicate with, others can be a powerful and transformative experience. Moreover, the very act of boldly taking over the airwaves can be as important as what is communicated and how widely it is disseminated. This book, then, is an examination of how some Canadian radio pirates have used unlicensed transmissions to occupy aural space, and to create meaning by fabricating islands of resistance out of thin air.

We, as media activists, reject the idea that radio is a hopelessly old-fashioned technology that is of no relevance in today's high-tech world. In response to the question of why one might choose to use radio wave transmissions when the internet has brought podcasting and web-streaming to our fingertips, we would respond that if pirate radio was simply about sharing sounds and ideas across a geographic space, then perhaps, yes, podcasting would be the preferred way to go. However, as Ron Sakolsky (Chapter 7) suggests, if pirate radio can be conceptualized as a form of "squatting the airwaves," then its potential challenge to the corporate state is rooted in the very nature of its unlawful existence, whereas podcasting offers no such direct confrontation with state or corporate authority and seems to be perfectly legal. Yet, if we look to the Volomedia patent granted by the US in 2009 for control of "the downloading of episodic media content" (i.e., podcasting), it has become clear that not even podcasting is safe from enclosure.[14] As Tetsuo Kogawa has pointed out, the age of satellite communications and the internet will only leave more room for radio to exist in micro-spaces. As he explains, "Change in a tiny space could resonate to larger space but without microscopic change no radical change would be possible."[15]

Sometimes, as is illustrated in Andrea Langlois and Gretchen King's account (Chapter 8) of the use of pirate radio in protest situations, pirates use both the airwaves and web streaming together. In one example they describe how this dual approach was effective in garnering a mutually-reinforcing array of international, national, and local support for the protest. Yet, in rural contexts, it is precisely the ubiquitous nature of the radio receiver that makes it the perfect tool. Most everyone has access to an inexpensive radio receiver but not necessarily to a computer, especially in the Global South. Such a widespread pattern of technological dispersion is not true of newer media technologies, and even if it should become so in the future, the decidedly localized impact of low power radio is typically considered a plus rather than a minus in pirate radio circles. In fact, pirate radio might best be viewed as one element within an ensemble of autonomous media.

As articulated by Christina Dunbar-Hester in her 2008 examination of gender, identity and activism with regards to low-power radio:

> Radio itself is viewed by some as a unique media technology, making access to it very appealing: radio does not require producers or listeners to be literate; it can reach a small, local community or area; production and broadcast technologies are relatively inexpensive and easy to use; radio is very inexpensive to receive; and it is easier and cheaper to provide programming in an aural-only medium than in a tele-visual one. In spite of charges of radio being a dead or dying medium, both activists and corporate broadcasters view the FM band as valuable.[16]

The importance of such accessibility is illustrated in Charles Mostoller's essay about Radio Barriere Lake (Chapter 6), an indigenous community where radios are a central feature of communication and the radio station serves to link members of the community together culturally. In a similar vein, Stephen Dunifer's essay (Chapter 2) on pirate radio activism among indigenous groups in Oaxaca serves to place the Canadian experience in an international context. Even with the advent of digital radio, we can assume that analog radio receivers will not disappear, but might become a liberating alternative, or even a counter-hegemonic force, in relation to the mass media model of communication. Some hope that in the move to digital radio the FM band will be abandoned by corporate and state interests. Others fear that it will instead be auctioned off to the highest bidder, leaving less space between stations than currently exists, which means that pirates will have difficulties finding empty spots on the FM dial.

Radio Art with a Pirate Twist

Historically, one of the longest running above ground Canadian pirate radio micro stations has been the 1-watt voice of Radio 90 narrowcasting from the Banff Centre's New Media Institute in Alberta. Because radios can be used as a mobile sound system, the airwaves can become an artistic tool for creating interactive spaces and experiences as Kathy Kennedy, Kristen Roos, Christof Migone, André Éric Létourneau, Anna Friz, Stephen Kelly and Eleanor King, and Marian van der Zon detail in their chapters. Kathy Kennedy (Chapter 16), for example, uses low-power radio to create concerts within public spaces or treasure hunts which are not about tuning in from home, but instead about how pirate radio can operate inclusively to allow audience participation and so transform the concept of performance accordingly.

Another possibility for artistic uses of pirate radio involves what some call "radio parties" or "festivals" in which the broadcast is conducted within a convivial atmosphere. Such parties can be political gatherings of large numbers of people aimed at occupying public space to create temporary autonomous zones similar to those events manifested by Reclaim the Streets, or they can involve more intimate and less overtly politicized settings. An example of the latter is illustrated by Anna Friz's Radio Free Parkdale (Chapter 13) events in Toronto. In these situations, she and her housemates held radio parties during which people could listen from home, or come to the party in person to participate in radio plays. In essence, then, one has the choice of listening or getting behind the microphone. Relying primarily on word of mouth within the neighbourhood to announce these upcoming gatherings, these radio parties — like the larger Temporary Autonomous Radio (TAR) music festivals described by Marian van der Zon (Chapter 9) or the smaller events hosted by Stephen Kelly and Eleanor King of Radio Ballroom Halifax (Chapter 15) — are more about the event itself and participation than about directly challenging access to public space or being concerned with the range of transmission and how many people might be listening at any given time.

When such parties take place in public spaces, transmitters and portable radios replace sound systems. Using such a do-it-yourself (DIY) approach, in August 2008, a Montreal group called the Pirates of the Lachine Canal threw a party where a 1-watt FM transmitter with a range of 100 feet was used to create a low-cost community event that could be easily moved if the police showed up. It did not require a large sound generator because the deejays plugged portable music devices into the transmitter's console and party-goers brought their own boom boxes. When Parks officials tried to shut down this self-organized block party because the Pirates did not have the permit required to hold a barbecue along the Lachine Canal, a local independent brewery loaned them their lawn and the party continued.[17] The officials made no mention of the illegality of the broadcast.

Policing the Airwaves

There are two bodies that regulate radio in Canada, the CRTC and Industry Canada. The CRTC deals with the "content, formatting and benefits to society"[8] of radio transmission, while Industry Canada oversees issues relating to the technical requirements of operating a radio station (wattage, interference, borders and use of the airwaves). The latter acts as the enforcement agency in cases of pirate radio transgression. For both regulatory bodies, any unlicensed radio in Canada is officially considered illegal. "The Radiocommunication Act stipulates that no radio apparatus that forms part of a broadcasting undertaking may be installed or operated without a broadcasting certificate issued by the Minister of Industry."[9] Any station seeking a broadcasting certificate or licence must apply for it. The granting of the licence is by no means certain, and the application process requires extensive documentation, a sizeable capital outlay and an agreement to abide by CRTC rules regarding studio construction, procedures, language and content.

Although unlicensed radio is deemed illegal in Canada, enforcement is typically complaint driven. Because Industry Canada is chronically short-staffed, this lack of enforcement agents on their part limits the active regulation of pirate radio. Sometimes Industry Canada staff discover a station at random and attempt to shut it down, but more commonly complaints are made by individuals who are offended by the content and/or language being broadcast, or else concerns are voiced by commercial radio stations claiming interference with their programming or reporting unlicensed broadcasting as a form of lawbreaking. Given the number of stations that we have become aware of which are, or have been, active in Canada over the years, it is fair to surmise that regulators do not view pirate radio as a high priority or more stations would have been shut down than has been the case. However, if Industry Canada representatives should knock on a radio pirate's door, there are a number of options available. It has been reported that these government enforcement agents typically ask that the pirate cease and desist — in other words, to quit broadcasting. If the station voluntarily complies, the Industry Canada personnel assigned to the case will likely leave it with only a warning. If voluntary compliance is not forthcoming, then steps may be taken to confiscate equipment and levy fines. Finally, if the pirate station insists on continuing to broadcast illegally, theoretically it may be subject to criminal and/or civil charges.[20]

It is difficult to ascertain how many pirate radio stations have been shut down historically. In a 2009 phone interview, a representative of Industry Canada stated that they did not have this information, and suggested that this lack of data is related to the dearth of pirate radio broadcasts occurring within the country.[21] Yet only a few months later, Radio Free Cortes in British Columbia received an official letter that threatened them with shutdown unless they applied for a licence. In a subsequent phone interview, a representative from the CRTC argued that because "Americans are more politicized"[22] in relation to radio than are Canadians, there are fewer pirate practitioners north of the border. In league with this stereotype of the apolitical Canadian who is quite satisfied with legally available radio options, Carla Brown, a journalist working for the CBC has stated, "Canada has an extensive community radio network that allows almost all types of content on the air. Canadian pirates tend to do it as a hobby rather than a political statement."[23] Despite a commonality of opinion between Industry Canada, the CRTC and CBC reporters like Brown, it has become clear to us in the course of compiling this volume that there are numerous and varied pirate radio practitioners across Canada. They are not merely hobbyists doing vanity broadcasting, but are choosing to take over the airwaves for a multitude of reasons, including political motivations.

An interesting case in point that reveals the sometimes highly politicized nature of Industry Canada's enforcement practices, occurred in February 2010 involving "Safe Assembly Radio," an unlicensed Vancouver low-watt station daring to broadcast views and opinions critical of the corporate and patriotic spectacle known as the Winter Olympics. While claiming to have immunity from the CRTC's rules and regulations on pirate radio because it was a temporary art-related project, combining its on-air broadcasts (which had a 3km radius) with internet streaming online from the artist-run VIVO Media Arts Centre (one of the few Vancouver-based cultural groups which refused to apply for Cultural Olympiad funding), after less than 24 hours of broadcast time the station was shut down by Industry Canada officers, who threatened VIVO as an organization with a fine of $25,000 a day and with fines of $5,000 per day for each individual involved in the broadcasts. Curiously dressed for the occasion in Vancouver 2010 jackets, the Industry Canada officials handed out business cards which sported a 2010 logo and e-mail address, and arrived to silence the radio dissenters in a Vancouver Olympics Organizing Committee (VANOC) vehicle.[24]

In spite of an undisclosed history of similar enforcement measures on the part of Industry Canada, some writers have nevertheless maintained that the reason that there have not been as many pirate radio stations in Canada as in the States is because the openness of the Canadian community radio sector makes piracy unnecessary. An example is Charles Fairchild's argument presented in the 1998 book *Seizing the Airwaves*. At that time, the FCC did not allow the issuance of low-power radio licences under 100 watts. Today, his arguments must be tempered by the existence of the FCC's low-power FM (LPFM) program which came into being at the turn of the century as a result of the rapid growth of the pirate radio movement in the United States. However, the new LPFM option was arguably initiated by the FCC as a divide and conquer strategy.[25] In relation to that movement, some radio pirates had merely wanted licences and so became advocates of LPFM, while others refused to submit to having their broadcasts legally sanctioned and wanted nothing to do with licensing. Taking into account what we consider to be the FCC's intentionally divisive goal for the LPFM service and the relatively small number of low-power licences actually granted by the program in comparison to the number initially promised, we can see that the mission, scope and impact of the US community radio sector is still quite limited in relation to access. However, with reference to licensed community radio outlets in Canada, Sheila Nopper (Chapter 4) describes the increasingly restrictive nature of stations, such as CIUT in Toronto, as a possible impetus for the potential flowering of Canadian pirate radio broadcasting in the future.

While the Canadian regulatory regime gives lip service to the positive weight allocated to social context and diversity of content in approving new applications for community radio licences, the middlebrow atmosphere of the CBC and the market demographics of the corporate sector are still what overwhelmingly characterize licensed programming. Moreover, their privileged position legally enables licensed stations to expand their broadcast range at the expense of new community radio applications. For example, in 2002, the proposed Gabriola Co-op Radio station, on Gabriola Island in British Columbia, began the process of applying for a low-power community licence. At the time of this writing, they are still involved in the application process.[26] Most recently, the delay was due to problems created by a subsequent application by Rogers Communications Inc. — which already operates a regional commercial station — to expand their broadcast range. In this situation, the Rogers application was given legal priority because they already held a licence as compared to the still unlicensed Gabriola station's earlier application. In 2000, a pirate station on Hornby Island, another Gulf Island in British Columbia, was visited by Industry Canada based on a complaint of illegal broadcasting and told to cease and desist. It complied and some of the programmers began the process of applying for a 5-watt developmental community licence. They are still mired in this application process as this book goes to press. Yet other former Free Radio Hornby pirates were never interested in being licensed in the first place and some have drifted to other pirate stations instead.

Conclusion

Pirate radio, by its illegal and often ephemeral nature, is difficult to document. In our research, we found a wide variety of stations plying the airwaves as pirates. We also met numerous dead ends, in large part because it is so often a clandestine activity. Since documenting a pirate station may lead to its discovery by regulators, we respect the decisions not to participate in this book that were made by those pirates who did not want to have their stations profiled here out of a concern about inadvertently increasing their vulnerability to shutdown. In other cases, we had leads, but were unable to find enough tangible details to weave them into a story. Consequently, despite our persistent inquiries, there are inevitably gaps within this anthology. In that sense, while this book maps previously uncharted waters, it is not definitive. We encourage others to fill in those gaps — from ethnic underground dance music stations in urban areas to rural stations actively engaged in cultivating regional life-ways. It is indeed possible that once this volume is published, more historical and present-day examples of pirate radio will start to surface like formerly submerged islands of radio resistance arising from the depths of the sea. Others, no doubt, will choose to remain invisible. One thing is certain — they will not cease to exist. After all, it is relatively easy to set up a pirate radio station. Since the late 1980s with the advent of micropower technology, a 50-watt radio transmitter can be as small in size as a loaf of bread. Moreover, a would-be pirate can now build her own transmitter very inexpensively, or, at a little more cost, order one online.[27]

In our editorial deliberations, we seriously questioned whether and how to document the stations and individuals that now appear within these pages, as it may bring them under the radar of Canada's regulatory bodies. In the end, we left it up to the individual practitioners themselves as to whether or not they wanted to be included here. Pirate stations remain elusive, and perhaps this is essential for their existence. If we found them, we welcomed their participation in our project. If they chose to engage with us, we included their stories so as to inspire others to become part of the pirate radio experience as readers, listeners, practitioners, artists, educators, activists and outlaws. Welcome aboard!

NOTES

1. That outsider status has been actively sought by some stations with permits or licences and by web-based practitioners. Historically, an example from the 1970s is the Calgary Cable FM station called Radio Radio, which existed for over 35 years, and had the reputation of being a pirate station, even though it did in fact have a permit. It did not broadcast over the airwaves — it cablecast — and had strong community support as an alternative radio station with a distinct pirate ambience.

2. Stevphen Shukaitis, "Dancing Amidst The Flames: Imagination and Self-Organization in a Minor Key," in *Subverting The Present, Imagining the Future: Insurrections, Movement, Commons*, ed. Werner Bonefield (New York: Autonomedia, 2008), 102.

3. Franco (Bifo) Berardi, *Felix Guattari: Thought, Friendship and Visionary Geography* (New York: Macmillan Publishers, 2008), 29–30.

4. Franco Berardi, M. Jacquenot and G. Vitali, *Ethereal Shadows: Communications and Power in Contemporary Italy*, (New York: Autonomedia, 2009), 80.

5. Michael Goddard, "Felix and Alice in Wonderland: The Encounter between Guattari and Berardi and the Post-Media Era," http://www.generation_online.org/p/fpbifo1.htm (accessed August 22, 2009).

6. A more thorough explanation of the relationship between wattage and distance of reception can be found in Chapter 9, "The Care and Feeding of Temporary Autonomous Radio" by Marian van der Zon.

7. For more information, see Anne MacLennan, Chapter 3.

8. Charles Fairchild, "The Canadian Alternative: A Brief History of Unlicensed and Low Power Radio" in *Seizing The Airwaves: A Free Radio Handbook*,

ed. Ron Sakolsky and Stephen Dunifer (San Francisco: AK Press, 1998), 50.

9. Ibid., 51.

10. Michael Keith, *Signals In The Air: Native Broadcasting in America* (Westport, Connecticut: Praeger Press, 1995), 88.

11. Stefano Harney, "Governance and the Undercommons," (2008), http://info.interactivist,net/node/10926 (accessed August 22, 2009).

12. Nick Couldry, "Being Elsewhere: The Politics and Methods of Researching Symbolic Exclusion," in *The Politics of Place*, ed. Cresswell, T. and Verstraete, G. (Amsterdam: University of Amsterdam/Rodopi Press, 2003), 109-124.

13. Ibid., 1.

14. Rebecca Jeschke, "EFF Tackles Bogus Podcasting Patent — And We Need Your Help," Electronic Frontier Foundation, November 19, 2009, http://www.eff.org/deeplinks2009/11/eff-tackles-bogus-podcasting-patent-and-we-need-you (accessed December 12, 2009).

15. Tetsuo Kogawa, "A Micro Radio Manifesto" (2006). http://anarchy.translocal.jp/radio/micro/ (accessed August 22, 2009).

16. Christina Dunbar-Hester, "Geeks, Meta-Geeks, and Gender Trouble: Activism, Identity and Low-power FM Radio," *Social Studies of Science*, 38-2 (April 2008), 203.

17. Rupert Bottenburg, "All Hands on Deck!" *Montreal Mirror*, 14 August 2008, 24, no. 9, http://www.montrealmirror.com/2008/081408/music1.html (accessed August 22, 2009). See also: www.myspace.com/piratesofthelachinecanal (accessed August 22, 2009).

18. Canadian Radio-television and Telecommunications Commission, website: http://www.crtc.gc.ca/eng/bctg-radio.htm (accessed August 22, 2009). There are limited exemptions where individuals can broadcast without a licence. See the CRTC's *Public Notice 2000-10* for more information. http://www.crtc.gc.ca/eng/archive/2000/PB2000-10.htm (accessed August 22, 2009).

19. Industry Canada, Spectrum Management and Telecommunications, "BPR (broadcasting procedures and rules) -1 General Rules," Issue 5 (January 2009), http://www.ic.gc.ca/eic/site/smt-gst.nsf/eng/sf01326.html (accessed August 22, 2009).

20. For more information, see the *Radiocommunication Act*, Section 9 (for criminal charges) and Section 18 (for civil charges), (15 June 2009), http://laws.justice.gc.ca/en/ShowFullDoc/cs/R-2//20090707/en (accessed August 22, 2009).

21. Personal telephone interview, Industry Canada representative, April 27, 2009.

22. Personal telephone interview, CRTC representative, April 27, 2009.

23. Carla Brown, "Pirate Radio: a Voice for the Disenfranchised," Peace and Environment News, July-August 1996, http://www.perc.ca/PEN/1996-07-08/s-brown.html (accessed August 22, 2009).

24. Dawn Paley, "VIVO Radio Signal Silenced by Industry Canada." Vancouver Media Co-op, http://vancouver.mediacoop.ca/story12769 (accessed March 13, 2010).

25. Ron Sakolsky, "The Myth of Government-Sponsored Revolution: A Cau-

tionary Tale," in *Creating Anarchy* (Liberty, Tennessee: Fifth Estate Books, 2005).

26. For more information on the Gabriola Co-op Radio process, see: http://members.shaw.ca/gabriolaradio/index.htm (accessed August 22, 2009).

27. There are many resources available for would-be pirates. Please see the Rip-Roaring Radical Radio References in this volume.

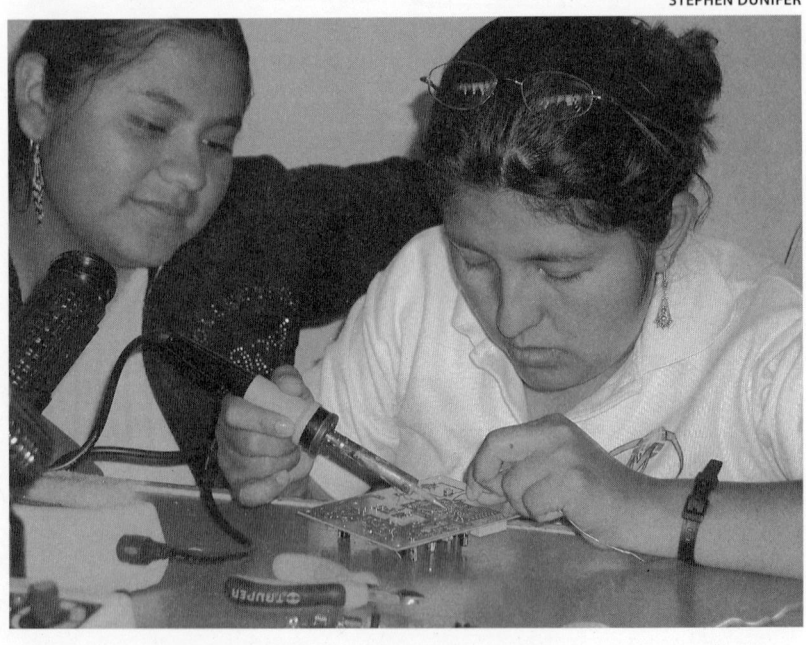

Students building a transmitter in a workshop led by Stephen Dunifer in Oaxaca, Mexico

CHAPTER 2

Latitudes of Rebellion
Free Radio in an International Context

Stephen Dunifer

IN THE INTERNATIONAL ARENA, "FREE RADIO" IS THE term best suited to describe the ongoing rebellion against not only control of the broadcast airwaves through licensure and sanctions, but the neo-liberal/free market paradigm as well. Entering the lexicon around the late 1960s, the term free radio was used to describe the broadcasting efforts of offshore broadcasters, such as Radio Caroline and Radio Veronica, operating in Europe. Popular support was widespread for these "pirate" broadcasters who played music and aired programming not heard on the BBC and other state controlled services. Even community radio as a broadcast form did not exist in Europe at that time, and is still somewhat limited. Although specific details are often difficult to obtain on the global breadth and depth of free radio broadcasting, the picture that emerges is one of a vibrant and universal movement. Unlike the rosters of community radio stations maintained by organizations like the World Association of Community Radio Broadcasters (AMARC), no central registry exists for free radio broadcast stations — due in large part to the elusive nature of the activity itself.

At its core free radio is an expression of immediacy and a rejection of state and corporate control. From very early on free radio has played a central role within popular struggles for liberation and self-determination internationally. Beginning in the late 1940s, Bolivian tin miners began to create radio stations as part of a larger process to counter ongoing repression by autocratic government and military forces. Over a period of 20 years, approximately 30 radio stations were established in the highland mining communities of Bolivia, most of them after the successful social uprising of 1952 that led to nationalization of the mines. Despite their ultimate destruction following the military coup of 1981, the legacy of these stations remains as one of the most outstanding examples of grassroots radio in history. Apparently this is still well understood in Bolivia where new community radio stations, now numbering about 30 with a goal of at least 50, are carrying on the already established tradition of street radio. During the indigenous protests that eventually culminated in the election of Evo Morales, street reporters and community radio stations played a vital role in maintaining and increasing the effectiveness of the protests, blockades and strikes. Unlike what might be termed NGO (non-governmental organization) radio, such grassroots radio stations do not originate under the auspices of a formal institution. Instead they arise from the participatory process of the community itself. As in the case of the Bolivian tin miners, and many other similar situations, free radio is a collective expression of the entire community. Full participation by the community is the heart of the radio station, not an afterthought or add-on as in the case of many so-called community radio stations.

Arising from the specific needs and issues of the community, free radio stations require no further legitimization other than that given by the communities creating them; outside legitimization is only a means by which to throttle expression, limit participation and stifle content. It is one thing to declare that free speech and uncensored communication are human rights as stated by the United Nations' (UN) Universal Declaration of Human Rights, but quite another struggle to act on these principles and assert control over the means of communication. Free speech, like other fundamental rights, is an inalienable right. It is as connected to human nature as breathing. Inalienable rights exist *a priori*; no institution or state can grant or confer them. Suppression, control and disregard, or protection and guardianship are the only options left to state and institutional actors.

Free radio has been integrated into a variety of popular struggles, from Radio Rebelde, established by Fidel Castro as part of the liberation of Cuba from the Batista regime, to Radio Venceremos in El

Salvador, and it has served as an important tool in the arsenal of the guerrilla forces fighting against the occupation of East Timor by Indonesia. It has become the voice of the favelas in Brazil where some 2000 free radio stations exist without government sanction or approval. When threatened with closure by government agencies, communities arise to defend their voices. Mass strike actions by taxi drivers forced the Taiwanese government to abandon its effort to shut down underground radio stations in the mid-1990s. On numerous occasions indigenous communities have put their bodies between their radio station and government forces attempting to shut them down. Following the Zapatista uprising in Mexico in 1994, a subsequent call was made by Subcomandante Marcos in 1996 for the creation of an international network of grassroots media. In response, independent community media then entered a new period of revitalization and regrowth in step with a burgeoning anti-globalization movement. Many community voices had been silenced not at the barrel of a gun, but by neo-liberal polices which privatized the broadcast airwaves and mandated their sale to the highest bidder. A single FM frequency or channel for the entire country of El Salvador had a price tag of $100,000. Onerous regulatory policies combined with civil and monetary sanctions were brought to bear against any community believing that they had the right of free speech and expression.

For those who resist, the steel fist of state-sanctioned police or military violence rests within the velvet glove of neo-liberalism and is enforced by a global corporate mafia. A handsome profit has been made in selling crowd suppression technology and weapons to both developing and first world countries as mass protests against corporate globalization and neo-liberal policies have broken out on an international scale. Close on the heels of the arms merchants came the lawyers and consultants representing private security firms and mercenaries. To avoid an embarrassing repetition of the shutdown of the World Trade Organization's (WTO) meetings in Seattle, steel cordons were raised in Genoa, Prague, Cancun, and dozens of other cities to protect the elite gatherings of the Group of Eight (G8) or WTO from the masses who were insisting that another world was possible. But, unlike people, radio waves cannot be easily fenced out. This is a primary reason why free radio is considered an ominous threat by those who wish to maintain their reign of domination and control. After all, the first paragraph in the *Dictatorship for Dummies* book states: "Seize the radio stations dummy." A slogan that evolved with Free Radio Berkeley goes like this: "If you cannot communicate, you cannot organize; if you cannot organize, you cannot fight back; and, if you cannot fight back, you have no hope of winning."

It may be difficult for people of First World media-saturated countries to understand the importance of free radio and community broadcasting to social movements abroad. For example, during the mid-1990s, a broadcast station was set up in the northern coastal farming area of Haiti. As part of a larger movement for land reform, this station began broadcasting what the market prices for crops should be in Creole, the native language. To many it is no big deal — just a farm report. For the farmers, however, it was the difference between barely making it and not making it at all. It was common practice for crop buyers to cheat the farmers by lying to them about the market prices. Without any means of knowing otherwise, the farmers undersold their crops. The deception came to a grinding halt when the farmers were informed of what the actual market prices were. Rich landowners and agricultural businessmen, threatened by these circumstances and increasing incidents of land seizure by the peasants, hired local police to destroy the radio station and kidnap its principal organizer, the mayor of the town, who was one of the leaders of the land reform movement. Despite the destruction of the station and the wounding of a night watchman, the mayor eluded capture. After the situation had calmed down a bit, the mayor demanded compensation from the government for the loss of the equipment and facility. Surprisingly, he eventually received it, enough to replace the equipment and even buy a more powerful transmitter. For some, radio is just entertainment, for others it is a lifeline.

Within the context of indigenous peoples throughout the Americas constituting themselves as one large community without borders and asserting their sovereignty, free radio and community broadcasting is construed as yet another sovereign right. With homes and villages destroyed by mud slides, rivers and lakes polluted, cancer rates off the charts, mountains ripped open and laid bare and forests stripped, indigenous people are all too well aware of their role as the canary in the coal mine of neo-liberal/free market fundamentalism. Moreover, free radio is a means by which they can preserve their languages, cultures and sovereignty. For indigenous people, the ability to communicate is a matter of life and death.

The Oaxaca Model

Nowhere has this struggle to communicate been more dramatically played out than recently within the Mexican state of Oaxaca. Mere coincidence cannot explain the fact that the poorest state in Mexico, Oaxaca, also has the highest percentage of indigenous people. During the early hours of June 14, 2006, 3000 state police armed with truncheons and shields carried out the order of Oaxacan Governor Ulises Ruiz Ortiz, to disperse the teachers union and the associated *Assemblea Popular de los Pueblos de Oaxaca* (APPO) and break up their *plantons* (encampments) in the City of Oaxaca. Special attention was paid to Radio Planton operating at the locus of the encampments in the *zócalo* (city centre), which was attacked and destroyed by the state police. This naked display of violence lit the fuse of resentment and rebellion on the part of indigenous communities who had been exploited and marginalized for generations. What began as an annual protest occupation by APPO in the capitol of Oaxaca quickly grew into a full-blown state of insurrection. Showing their resolve, the teachers and their community supporters retook the *zócalo* after the police retreated.

Radio Planton, originally conceived in 1998 by the teachers union, began its first broadcast in the city of Oaxaca on the morning of May 23, 2005 at 94.1 MHz as a voice for not only the teachers but for the community as a whole. It quickly became broadly reflective of the diverse aspects and nature of Oaxacan society with 70 percent of the programming being representative of that larger community. After the attack of June 14 the local university's two radio stations, one FM and the other AM, became the voice of the teachers and community — Radio Universidad. Responding to broadcasts on Radio Universidad for massive nonviolent civil disobedience, virtually all government buildings in the City of Oaxaca were shut down by either occupations or blockades. Constructed of everything from bricks to burned-out cars and buses, barricades appeared on every major street. Government city halls and other buildings were taken over in 25 other towns as well. Thus began what was to be called the Oaxaca Commune. On August 1, 2006, two thousand women marched to the state TV Channel 9 facility to demand an hour of airtime so that their truth would be told. Rebuffed but not stymied, the women took over the facility which included one FM and one AM station as well. In so doing, they wrote another chapter in the history of people, and most particularly women, seizing the means of communications and reclaiming what is theirs. By evening the women were broadcasting on Channel 9 with demands for the resignation of the governor. Videos by indigenous community members followed the initial broadcast. For the next approximately three weeks, the indigenous communities saw what they have never seen before on Channel 9 — themselves!

In response to the occupation, armed paramilitaries and police attacked the main transmitter and support equipment for Channel 9 in the early morning hours of August 21. High velocity bullets ripped into equipment, effectively putting Channel 9 off the air. One person was wounded. As the word spread about this attack, a spontaneous movement seized 12 to 15 commercial radio stations in Oaxaca City. Expecting to be attacked at any time, neighbourhoods and communities throughout Oaxaca City organized a complex network of barricades and notification systems, using materials such as bells or fireworks to warn of an impending attack by the police and/or paramilitaries. The people were in control and the official government no longer functioned in many parts of the state of Oaxaca.

Humiliated by the turn of events, the governor and his allies in both the Mexican government and private sector commenced a "dirty war" against the popular assembly movement. Reminiscent of similar tactics employed in Central America in the 1980s, people were "disappeared" and became targets of "random" shootings. One of the victims of this "dirty war" was Brad Will, an American journalist with Indymedia and a documentary filmmaker. He was shot and killed on October 27, 2006 by police and paramilitaries acting on behalf of the governor. Interestingly enough, Brad had been involved in the creation of a free radio station, Steal This Radio, in New York City in the mid-1990s. Increasing numbers of federal troops were brought in to crush the popular rebellion. Finally, a force numbering approximately 4000 were dispatched in November 2006 to recapture Oaxaca City and return it to "normalcy." Despite repeated attacks, including being strafed with bullets, Radio Universidad continued to broadcast until the very end as the voice of the Oaxaca Commune. Police forces were never able to invade and shut down the station. Fierce and determined resistance prevented federal police from entering the university. Free radio stations were operating in other communities as well. Trying to copy the radio efforts of the popular assembly movement, the political party of the governor, the *Partido Revolucionario Institucional*, put its own station on the air as part of a disinformation campaign.

It is impossible to properly cover all aspects of what transpired in Oaxaca during this period within the context of this chapter. Although widely covered by independent media outlets and progressively-oriented Mexican newspapers such as *La Jornada*, mainstream sources both outside of and inside Mexico were virtually silent. When they did choose to speak, it was to blame the popular movement for the violence and to provide cover for the actions of the governor and the police. In Mexico, the television broadcast media outlets are controlled by only two entities — Telvisa and Azteca. Being both pro-corporate and pro-government, neither entity will ever speak truth to power. They are content in their roles as stenographers for the elite and continue their efforts to pacify the population with a plethora of mindless entertainment. Although the dominant population of Mexico is indigenous, they are rarely seen or heard in the established Mexican media. When they do make an appearance, it is usually to be portrayed in a negative light. A rigid caste system has existed in Mexico since the arrival of the Spanish colonizers. This underscores the importance of what has transpired with the popular assembly movement in Oaxaca and why free radio stations in the hands of indigenous communities are a vital part of the ongoing struggle for self-determination and freedom. Their narrative cannot simply be fit into a preordained leftist mould.

Overall, media policy in Mexico is in a rather retrograde position when compared to other countries in Latin America — even Columbia saw the necessity for community radio and recently issued hundreds of blanket licences. Community radio had essentially been considered illegal until new legislation made some provisions to legalize it. Expectedly, most existing community radio stations were not invited to the table to discuss the provisions of this legislation. While the South American division of AMARC has a progressive and radical history, the same cannot be said of the Mexican branch. Instead, the Mexican representative of AMARC appointed a handpicked group of delegates from a small number of stations and interjected herself into the process. This onerous arrangement resulted in about half of 15 community radio stations being shut down as part of the deal, another indicator of why not to trust NGO representatives indiscriminately, as they can have their own agendas and self-promotion as their primary operating principles.

It is clear that both indigenous communities and popular assemblies and movements in Mexico, and Oaxaca specifically, have not been waiting for legitimization by any entity, government or otherwise. In Oaxaca, as of May 2009, there were dozens of free community radio stations on the air, with 150-200 stations operating in the entire country. Radio Planton returned to the air in early 2007. When you ask these communities about the importance of their radio stations, some common themes emerge. They are means by which to preserve language and culture, to bring news and information to the community, to organize against further exploitation and stealing of resources, to empower women and children to exercise their voices, and to entertain with music and stories. Because of their power, there are various actors who will kill to silence them. Two women working with Radio Copala, the voice of the Triqui community of San Juan Copala, were murdered on April 7, 2008, by seven gunmen wielding AK-47s. Their car was ambushed while they were on their way to a community radio workshop in Oaxaca City. Two other people in the car were injured and a four-year-old child barely escaped harm. Mexico is one of the most dangerous places in the world for journalists and media activists.

Based on my own personal experiences in conducting transmitter building workshops in Mexico, there is both a pressing need and demand to establish more free radio stations, not only in Mexico but throughout the world. Primarily, the obstacles to an even more vigourous growth of community broadcasting are funding, training and support. In January 2007, Free Radio Berkeley's Project TUPA (Transmitters Uniting the Peoples of the Americas) in conjunction with local organizations and people conducted two five-day transmitter building and radio station creation workshops in Oaxaca City. Attended by about 50 people, mostly in their early twenties (some younger, some a bit older), who represented 24 Oaxacan communities, these technical workshops were accompanied by evening sessions on the social aspects of community radio and provided the represented communities with the knowledge and equipment to establish their own radio stations. With small grants and personal donations totalling around US$12,000 to $14,000 to cover equipment costs and operational expenses, such as food and rental of facilities, these workshops proved to be very cost effective — 24 radio stations for an average cost of about US$600 per station. As proven in Oaxaca, radio has an immediacy and flexibility that no other medium possesses. All you need is a transmitter, a properly situated antenna, a mixer, one or two microphones and a CD or mp3 player. Put everything on a table, make your connections, position the antenna and go on the air within 15 minutes. Anyone within range with a radio is a potential listener. Some have suggested that radio is no longer necessary now that we have the internet. Such a view is dangerously naïve. Sever a few critical fibre optical cables and there goes the network. Further, it is very First World-centric. For the equivalent cost of one or two computers (US$1,000–$1,500), a complete radio station covering a radius of eight to ten miles can be established.

Future Directions in Technology

As the June 2009 social unrest in Iran underscored, not enough emphasis can be placed on the necessity of having a decentralized means of communications. With the digerati extolling the role and impact of social networking sites, cell phones, and personal digital assistants (PDA) on the ongoing protests in Iran, an obvious weakness of these centralized networks has been exposed for all of those who care to examine it. Iran's communications network, installed by a joint venture of Nokia and Siemens, came with a monitoring centre whose capabilities include the examination and control of every byte of data passing through it. A process called deep packet inspection allows for the ability to troll for keywords and block any communications containing those words. This is far more insidious and effective than merely blocking specific internet site locations which are assigned a unique address known as an IP address. IP address blocking can be countered by the deploying of proxy servers with constantly changing IP addresses, an activity cyber-activists have been engaged in to support the protests in Iran. Further, most cell phones now come with global positioning system (GPS) receivers, which allow for the user's location to be immediately known whether the cell phone is turned off or on. Older model cell phones can be tracked by tower triangulation. Software programs can be downloaded on cell phones to turn them into monitoring devices for any conversations taking place within the range of the microphone, all without the permission or awareness of the user. Such technologies may be much more Faustian than utopian, especially in light of programs such as Echelon and the installation of FBI black box taps (known as Carnivore) on the servers of every internet service provider.

Within this specific context, free radio becomes all the more important because it cannot be centrally controlled and shut down. Every tool has both strengths and limitations and any intelligent user of media tools must recognize this fact. Reliance on any one tool is foolish and short-sighted. Further innovations must be created and established to put technology to work for people and communities. Cory Doctorow, in his sci-fi novel, *Little Brother*,[1] shows a possible way forward with the development of "extranets" — local wireless mesh networks that allow for regional and local communications. Created with inexpensive wifi transceivers and software for self-configuration, extranets would be a way for local control of communications to be exerted. Software-defined radio receivers are yet another emerging possibility.

It cannot be denied that the internet has made the world much smaller, allowing information and news to flow in ways unimaginable a decade ago. Equally important to consider though, is that information without context is propaganda. De-contextualization is a primary means of control. Free radio is able to provide context in an immediate and direct manner. As part of a synergistic deployment of media and communications controlled by people, not corporations and government, free radio is a plant which only needs further watering and propagation to maximize its inherent possibilities. Let a thousand transmitters blossom!

NOTES

1. Cory Doctorow, *Little Brother* (New York: Macmillan/Tor-Forge Books, 2008).

DUNCAN MACPHAIL

Women on-air, World War II

CHAPTER 3

Resistance to Regulation Among Early Canadian Broadcasters and Listeners

Anne F. MacLennan

RADIO PIRACY DEPENDS ON ACTS OF TRANSGRESSION that in turn are bureaucratically defined by regulation. The slow development of radio regulation in Canada meant that what would now be considered piracy, at least initially, was sporadic and inconsistent. While pirate radio was distinct from amateur radio, it was almost impossible to break the rules until licences were issued for commercial broadcasting in 1922. Over the course of its first few decades, piracy assumed many forms in Canadian broadcasting. As radio became licensed, pirates — and even some of those stations who chose to get licences — challenged regulation in a variety of ways. Forms of resistance encompassed program content, frequency radius and listener responses to licensing, which all could be positioned just off the margins of acceptable broadcasting or listening practices.

Camping on the Airwaves in the 1920s

Initial incidences of piracy did not refer to pirate stations on unassigned frequencies, but strictly to the pirating of wavelengths. The transgressors in this case tended to be American broadcasters, occupying a space on the dial assigned to a Canadian radio station. This situation was illustrated in a 1927 headline proclaiming, "American Radio Stations Pirating on CKY, Lowry." These American stations "camped," or "pirated," the few wavelengths available to Canadian broadcasters making it difficult or impossible for listeners to hear their local Winnipeg station, CKY. Interference was one of the major issues in the 1920s and 1930s. High-powered radio stations in the United States regularly broadcast to large regions that did not respect the boundary between Canada and the United States. Some American radio stations were unwilling to work within the confines of their own regulations and circumvented American broadcasting norms by moving to Mexico so that they could broadcast into the United States, and, sometimes with the aid of very powerful signals, to Canada. Policing American signals was an early concern, but eventually regulation expanded to include Canadian stations as well.

Limited Regulation in Radio's Early Days

In fact, the lack of early regulation of radio broadcasting contributed to the tendency toward pirate activities. Canada's system is typically depicted as a hybrid because it incorporates private and public broadcasting. In the early stages of Canadian radio, decisions about the nature of public broadcasting were delayed. This situation was unlike that in the United Kingdom, where the General Post Office interpreted broadcasting as part of its mandate by forming the British Broadcasting Company in 1922. Under Royal Charter on December 31, 1926, it became the British Broadcasting Corporation in order to provide a national public broadcasting service. In 1912, the Radio Act in the United States accorded regulatory powers over radio to the Secretary of Commerce and Labor, making radio a commercial and private enterprise. Regulation of Canadian radio emerged in stages. A public network came into being with the Radio Broadcasting Act of 1932, which set up the Canadian Radio Broadcasting Commission (CRBC), and, in 1936, the Canadian Broadcasting Corporation (CBC) was created.

The regulation of wireless communications in Canada began with the Wireless Telegraphy Act in 1905 designating the Radio Branch of the Department of Marine and Fisheries as the licensing body. The Act was replaced with the Radiotelegraph Act of 1913, which granted the federal government control over all aspects of radio, including broadcasting. The Radio Branch continued its work under the Department of Naval Service from 1913 until 1922, when the department was abolished, and regulatory powers were returned to the Department of Marine and Fisheries. At first all licences were experimental; then in 1922 they were divided into broadcasting and receiving licences.[2]

Back in the early years of radio, however, ship-to-shore communication and telegraphy had a role to play in establishing regulatory patterns. The availability of ships' radio equipment provided a source of small transmitters and accounted for a degree of mobility for some stations. Commercial broadcasting was suspended during the First World War, but hobbyists returned from the war with training in radio communications. These skills fuelled the growth of the hobby and eventually the industry. Few listeners could afford an expensive floor model radio in the 1920s, so the majority of them assembled their own inexpensive crystal sets from easily available materials. Early listeners on crystal sets needed earphones so radio started largely as a solitary activity, usually for boys and young men, in barns, garages or attics to accommodate the leaky tubes. Women were also active in amateur radio for many of the same reasons as their male counterparts.[3] Hobbyists not only listened for entertainment, but "DXing," or scanning the airwaves of distant stations, consumed much of their time. Great distance was an achievement confirmed by postcards and recorded on maps by DXers.

Resisting Frequency and Content Regulation:
Religion, Foreign Languages and Politics

In the 1920s the first roster of Canadian broadcasters was almost all independent and private. Many of the early stations were owned and operated by those who had interests in newspapers, manufacturing radios and railway companies. At the other end of the spectrum, religious and other small but devoted broadcasters often valued their messages over regulation. Dennis J. Duffy recounts a violation of the regulations concerning the assignment of frequencies, in *Imagine Please: Early Radio Broadcasting in British Columbia*.[4] Duffy recounts how Dr. Clem Davies of the Centennial Methodist Church had his radio station in operation in time for Easter Sunday of 1923. By 1924 Dr. Davies moved his ministry to the Victoria City Temple and moved the radio station along with him by relocating the radio transmitter to downtown Victoria and shifting position on the dial to CFCT.[5] He initially operated without a licence as an amateur, not waiting for official sanction, and then moved onto another licensed radio station before any penalty could be considered. Davies' move was symptomatic of how easily broadcasters could set up a station with limited equipment in the early twentieth century. The poorly financed and independent stations that were prevalent in Canada throughout the 1920s and 1930s were less likely to suffer serious penalties. The key to the continued existence of such stations was the lack of complaints from the audience or, in a more positive sense, community support.

Resistance to broadcaster regulation was not confined to the relocation of stations. Some were simply never assigned commercial licences. Nadine Kozak has documented the persistence of the amateur station 10AB in Moose Jaw, Saskatchewan, from 1922 to 1934 despite its inability to obtain a commercial licence. The station was supported by local merchants demonstrating not only the tolerance of radio broadcasting outside of the framework of officially sanctioned federally licensed stations, but also approval of its existence. As the medium developed and what was seen as acceptable content and use became more fully defined, the possibilities of transgression increased accordingly. Many radio supporters viewed legislation of all types to be a hindrance to the continued operation of radio stations. As reported by the *Vancouver Evening Sun* in 1926, legislation, such as the amendment contemplated to require radio stations "to pay an indefinite royalty on copyright music," imposed hardship on small broadcasters.[6] The *Sun* writer argued that stations were not "revenue-producing" and the royalties would require the addition of "worthless" advertising. The *Sun* also made the case that radio stimulated the sale of copyright music, concluding that "[r]adio is an educational and cultural force. Its future should not be jeopardized by legislation that can serve no legitimate purpose."[7] This was particularly true in the case of Vancouver, where no American affiliate took root and many small stations shared a single frequency, until the arrival of the CRBC and CBC stations.

Beyond music, the regulation of Canadian radio also focused on spoken word programming, and was sometimes applied to language, religion and politics. During the Second World War, despite a ban on languages other than French and English, foreign languages were frequently heard — particularly in the West, where there were large populations of Canadians with other European mother-tongues. These broadcasts often resulted in complaints, although some uses of foreign languages were out of habit and perhaps not a direct or intentional violation of the ban. Yet some accused the government of initiating a witch hunt, particularly with regard to the German language during the Second World War.[8] A letter to L.W. Brockington consisted of a request to lift the ban on German-language broadcasts asserting that many of the 500,000 German Canadians, who spoke German almost exclusively, would benefit from Canadian German-language broadcasts to counteract the "damage done by German shortwave stations."[9] This "damage" acknowledged the prolific use of Nazi propaganda broadcasts over German airwaves, within the country and internationally through the use of shortwave radio.

Just as potentially disloyal commentary in languages other than English and French were not allowed, sentiments about religion ran high in the 1930s, and radio provided a platform for the expression of a variety of opinions. For example, Rev. Morris Ziedman of the Protestant Radio League took to the airwaves over CFRB in Toronto on November 29, 1936, to protest the Separate School Tax Bill. In his address he noted the many failings of the provincial government, but specifically focused on "the subject of State subsidy of the Pope's Church, which is of such vital importance to us as Protestants, because our spiritual forefathers fought, were tortured, and died for freedom of conscience, and Protestant Faith."[10] He goes on in his prepared text to say, "It is because Protestants are being tantalized, irritated and razzed by a minority," and in a nod to the imagined power of the censor, "I had better not say what I was going to say, because I might be put off the air — I had better say, by non-Protestants."[11] That comment, consistent with the rest of the talk, conveys the sense that divergent views would be censored and not tolerated over the air. Rev. Ziedman's speech would have been submitted to the CBC as a prepared text in advance, as was required by regulation. With this in mind, Rev. Ziedman's comment on what he "had better not say," indicates that he may have been capitalizing on the feeling that radio talks were live and that the comment may have been perceived to have been extemporaneous within the context of previous conflict about broadcasting religious commentary. Eventually the suspension of Rev. Morris Ziedman's religious programs by the CBC occurred due to the contravention of the regulation prohibiting abusive comments about any religious group.[12] Complaints directed the attention of CBC to every denomination as they each, in turn, were offended by the comments of other religious groups.

Political statements were equally as likely to garner complaints and influence decisions over what material was suitable to be broadcast. For example, CKGB in Timmins, Ontario, had allowed the Communist Party to broadcast as a part of its schedule, taking advantage of the leeway allowed by the CBC for local stations to approve political speeches. As noted in a letter from Thos. Lawrence, Secretary of the Northern Ontario Regional Committee of the Communist Party of Canada, to Gladstone Murray of the CBC, CKGB's station manager, Mr. W. Wren withdrew permission and terminated a previous agreement to allow the Communist Party on the air. Lawrence noted that Wren explained, "It is difficult to interpret the rules and regulations of the CBC particularly in regard to political speeches. Furthermore, the onus for the passing of such speeches rests on the local station ... We have received complaints from certain people regarding Communist broadcasts."[13] In this case, complaints from listeners provided the impetus or excuse for change despite the allowances in policy.

Conflicts over content, local priorities, language and the ability to conform to national network radio policies continued to spark resistance in every region of Canada. There were ongoing disputes over language — starting with complaints about French-language broadcasts in the West over CRBC and English in Quebec, as well as those language bans during the Second World War mentioned above. Local and regional priorities, such as those in Moose Jaw, also continued to mark the Canadian experience of radio. Once broadcasting patterns were formed in the Canadian South, and CBC network radio and television were expanded to the North through the Anik satellite in the 1970s, new forms of resistance arose. Illegal stations, unauthorized broadcasts in Inuktitut, broadcasts from Radio Moscow and other makeshift arrangements to meet local needs defied regulation, even co-opting some CBC network staff in the process.[14]

Listener Resistance

The implementation of licensing was not relegated to the stations themselves, but was also extended to receivers. A little known part of Canadian radio history is that the Radiotelegraph Act also required listeners who owned a radio set to pay a fee of one dollar a year to listen (broadcasters paid 50 dollars for their licences). The radio receiving licence fee was raised to two dollars in 1932 and $2.50 in 1938 until it was dropped completely in 1953. However, before the fee was eliminated, a lack of compliance with regulation in Canada extended to listeners as well as broadcasters. As radio grew, so did its licensed radio receiver sets from 9,956 in the fiscal year ending March 31, 1923, to 297,398 by 1929.[15] Canadians were known to resent and evade the payment of a licence fee, partly because Americans relied on privately sponsored broadcasters and had no licence fees. The belief that radio should be "free" persisted, especially since the majority of the population lived close enough to the border to receive American radio signals, for which no paid licences were required.

This early resentment might, in part, account for the continued critical view of the CBC and its broadcasts today in some quarters. Although the unwillingness to comply with licensing regulations can be compared with contemporary efforts to surreptitiously obtain cable television without payment, the phenomenon fits more comfortably into the mindset of the 1930s. Canadian census data in 1931 indicated that, for the first time, the urban population outnumbered the rural population (with urban communities defined as those greater than 30,000). Living in sparsely populated areas, Canadians were unaccustomed to the many forms of regulation that increasingly expanded in the twentieth century. The efforts to thwart radio inspectors in their attempts to collect fees for radio receiving sets parallel the simultaneous efforts of Canadian rum runners and moonshiners that frequently hid their equipment in the woods or on farms. Prohibition — the ban of the sale, manufacture and transportation of alcohol for consumption — extended from 1919 to 1933 nationally in the United States and for variable time periods in each Canadian province during the first half of the century. In fact, Toronto's CKGW, owned and operated by Gooderham and Worts Distillery, maintained a presence over the airwaves for its products, legitimately marketed in Ontario, but was still largely banned in the vicinity of the station's target audience immediately South of CKGW in the United States, where its strong signal could undoubtedly be heard. As an NBC affiliate, CKGW became the most active Canadian radio station in the North American broadcasting environment.

Typically, densely populated countries report incidences of shared listening in pubs or other communal centres. However, the vast rural nature of the greater part of Canada in the early twentieth century made this unlikely and at times impossible. Licence fees applied to each radio receiving set, which became a contentious issue with the rapid spread of radio and the ownership of more than one set. Shared listening occurred more often in homes and among family members. Those without radios were acutely aware of the sounds of radio that wafted out of their neighbour's windows in the summer months. The ability to evade licence fees in small communities therefore depended on the privacy afforded by isolation and the clannishness of community that protected the members from the perceived interference of outsiders, such as radio inspectors.

Being advertiser-supported and not requiring listener licences, American network broadcasting filled in the gaps, and provided competition to Canadian public broadcasting. As late as 1937, stories and editorials such as "Radio Licence Worries" in *The Montreal Gazette*, persisted in expressing the annoyance householders felt about the radio inspectors, who visited homes to check whether or not the household possessed a valid radio licence.[16] The suspicion of and resistance to radio licences was reinforced by the fraudulent sale of licences in Saint John, New Brunswick.[17] The sale of more than a million radio listening permits in Toronto was lauded, but many fines were anticipated as 650 homes with receivers that had been licensed in 1935 did not purchase permits in 1936.[18] While listeners typically resented the very existence of licence fees, the price increase in 1938 provoked a renewed outcry against them. In that same year, 1002 complaints were recorded in an internal CBC memo. The Department of Transport received 861 of the complaints and the remaining 141 went straight to the CBC.[19] The same file also included a 22-page petition resisting the licence fee increase.

A newspaper editorial, entitled "Canadian Broadcasting System — Or Is It?" outlined most of the protests against the final increase of the licence fee to $2.50 in 1939. It quoted Alex Frost, chairman of the radio committee, saying, "[n]ot only are we paying twice now, but will be paying three or four times over according to the number of sets owned, in the home, the car or summer camp."[20] The editorial went on to argue that "Radio subscribers... want to know if privately-owned broadcasting stations are riding free on the CBC system's corporation programs... why the Canadian system should be loaded up with commercials, and then linked up with the American hookup for more commercials... and would like to know why Canada cannot have a CBC program free of commercial blabla entirely?"[21] The resentment over licence fees became inextricably linked to the CBC, and popular perceptions of its role and potential as a publicly supported network dependent on those fees. Other political commentators were incensed over the hardship inflicted upon the poor by the licence fee during the Depression of the 1930s.[22] Because the CBC had held the role of national network broadcaster and regulator of both private and public broadcasting across the country since 1936, it was placed in a position of eternal conflict of interest unable to satisfy critics on all sides. Licence fees both supported operations and fed discontent with every aspect of Canadian radio.[23]

Sharing Frequencies

Another issue of concern to both listeners and broadcasters was interference. By 1930, the Radio Branch of the Department of Marine spent $250,000 annually to suppress what they deemed to be interference.[24] Problems with interference were generally solved by the sharing of frequencies, with exceptions in larger cities. Most stations across the country did not broadcast full-time and shared frequencies as well as equipment and/or studios. Generally, early Canadian radio broadcasters were not very well funded and could not have used the entire broadcast day had it been available. Canadian cities were not large enough to support many stations, so few conflicts arose over frequency sharing. The assignment of shared frequencies, prior to 1933, favoured listeners because so many of them still used crystal sets, which made tuning more difficult. The greater radio coverage of the country by 1939, when the CBC opened its last two powerful regional radio stations in Watrous, Saskatchewan, and Sackville, New Brunswick, made radio reception better, and a choice of more than one station feasible in areas distant from the American border.

However, outside interference from American and Mexican stations remained a problem for Canadian regulators. On December 12, 1935, K.A. MacKinnon of the Engineering Division submitted "A Report on the Interference from Mexican Stations on the 840, 910 and 960 Kc. Channels," and noted that his program monitoring at Ottawa and Strathburn, Ontario, indicated the interference "between CRCT [a Toronto CRBC station] and the Mexican-based station XERA was sufficiently strong to ruin any listener's enjoyment. In general XERA was the stronger."[25] MacKinnon also examined interference between XENT and CRCM (a Montréal CRBC station), and between CKY (a Winnipeg CRBC affiliate station) and XEAW. He concluded that the interference in these cases was minor and within the "assumed service range."

It is important to note that this equation of pirate radio with interference was about radio broadcasts that were usually directed at capturing a larger audience in the United States while avoiding US regulations. XENT in Nuevo Laredo, XEAW in Reynosa, and XERA in Villa Acuna, called "border blasters," migrated to Mexico from the United States. The stations, however, did not broadcast in Spanish, but in English. For instance, XERA was run by Dr. Brinkley, who gained fame for his goat-gland surgeries to treat "lagging libido." His treatments started in Milford, Kansas during the 1920s and his "goat-gland gospel" was broadcast over his radio station KFKB, one of the first in the American Midwest.[26] After an increase in the radio station's power, 3000 letters came in a day inquiring about cures, until Dr. Brinkley came under attack when Dr. Morris Fishbein initiated a campaign to revoke Brinkley's licence to practice medicine.[27]

Despite the fact that the station was operating under a cloud due to a US Federal Radio Commission (FRC) investigation to potentially rescind its licence, KFKB was "voted the most popular radio station in America in a survey conducted by the Chicago-based *Radio Times*."[28] Brinkley sold KFKB and started operations in Villa Acuna, Mexico with greater power as XER in 1932 and XERA by 1935, following the example of the "nation's station" WLW, Cincinnati, Ohio. WLW was granted sufficient power by the Federal Radio Commission to be heard across the United States, but its directional antenna moderated the strength of its signal in Canada. Similar to XERA however, XEAW and XENT, mentioned in MacKinnon's report, and other stations, fled the regulation of the FRC for Mexico where they could broadcast to most of the United States and parts of Canada. Radio personality Wolfman Jack credits his career to border radio as featured in the 1973 film, *American Graffiti*.[29] The long-standing practice of border radio just south of the American frontier only ceased to interfere with Canadian radio stations after the Havana Treaty of 1937, when "tolerable" levels of radio interference by sky waves after dusk were determined for North America.

Within Canada, interference drew the attention of regulators to both legal and illegal stations. Attempting to enforce public standards of morality, police seized thousands of dollars worth of radio equipment in raids when they closed down six radio transmitters that were operating illegally in Montréal after receiving complaints about interference with police radio and the broadcasting of obscene programming.[30] For regulators, the overall goal was to shut down any such pirate or unlicensed stations. One of the primary reasons for the regulation of these stations was the assumption by regulators that there was a finite number of North American frequencies to be allocated. Using the logic of inevitable scarcity, the existence of interference was treated as being the result of rogue broadcasts rather than the unequal distribution of access to the airwaves and the relatively large size of corporate stations, both private and public, in comparison to low-power radio, and especially low-watt pirate stations.

Conclusion

Resistance to regulation by early broadcasters was tolerated by the Radio Department of the Minister of Marine unless listeners or larger stations complained about the content of programming or interference. Prior to the formation of a national network, many deviations from the expected course were ignored or accepted. The stations that did not conform to the rules were generally small, low-powered stations serving their local communities. Even when taken as a whole, early Canadian radio was characterized by a constant stretching and testing of the developing regulation, making piracy or transgression common. Radio practitioners, including pirates as well as many listeners, challenged what they contended were the artificial limitations on access afforded by the Canadian regulatory regime and ignored or resisted their control. In doing so, they rejected the regulator's idea that Canadian radio broadcasters and listeners should gratefully embrace the Canadian hybrid — a public broadcasting system similar to the BBC that simultaneously reaped the benefits of an advertiser-supported commercial system like that in the United States. Because of the historical collision of viewpoints on how radio might be used in a Canadian context, it was only regulated once practice was already established, which guaranteed that breaking the rules would continue indefinitely as a feature of Canadian radio.

NOTES

1. "American Radio Station Pirating on CKY, Lowry," newspaper clipping, circa 1927, National Archives of Canada, RG 12, vol. 864, file 6206-162, vol.1.

2. In the first issuance of commercial radio broadcasting licences in 1923, 34 licences were issued. By 1957 there were 203 AM stations, 26 FM stations and eight short-wave stations and broadcasters were required to pay a fee based on their gross revenue. There were no further changes to the regulation of Canadian radio until the Radio Broadcasting Act of 1932 under the Conservative government of Prime Minister Richard B. Bennett (Canada, Dominion Bureau of Statistics, Information Services Division, Canada Year Book Section, *The Canada Year Book 1957-58* (Ottawa: Queen's Printer, 1958), 907.

3. Michele Hilmes uses an extensive list of articles from QST, the journal of the American Radio Relay League, and suggests that the "ability to escape the determinations of gender" may have added to radio's appeal. Michele Hilmes, *Radio Voices: American Broadcasting, 1922-1952* (Minneapolis: University of Minnesota Press, 1997), 132-136.

4. Dennis J Duffy, *Early Radio Broadcasting in British Columbia* (Victoria: Provincial Archives of British Columbia, 1983), 24-25.

5. Ibid.

6. "Radio and Copyright Music," *Vancouver Evening Sun*, January 29, 1926, National Archives of Canada, RG 33, 14, vol. 4.

7. Ibid.

8. Letter from A. H. Brinkman to the Director of the CBC, Craigmyle, Alberta, December 25, 1939, National Archives of Canada, RG 41, vol. 154, file 9-34 (pt. 1).

9. Letter from W. Madeley Crichton, Crichton, McClure & Co., Barristers, Solicitors etc., to L.W. Brockington (CBC), September 29, 1939; letter from W. Madeley Crichton, Crichton, McClure & Co., Barristers, Solicitors etc., to L.W. Brockington (CBC), October 2, 1939; letter from W. Madeley Crichton, Crichton, McClure & Co., Barristers, Solicitors etc., to Gladstone Murray (General Manager, CBC), October 6, 1939, National Archives of Canada, RG 41, vol. 154, file 9–34 (pt. 1).

10. Rev. Morris Ziedman, "Where There is No Vision the People Perish, " Protestant Radio League Broadcast, CFRB, Sunday, November 29, 1936 1:30-1:45 PM, National Archives of Canada, RG 41, vol. 40, file 2-2-8-2 (pt. 4), 2.

11. Ibid.

12. Appendix to letter from H.R.L. Henry, Private Secretary, Office of the Prime Minister to Gladstone Murray (General Manager, CBC), March 24, 1939, National Archives of Canada, RG 41, vol. 393, vol. 21-9.

13. Letter from Thos. Lawrence, Secretary, Northern Ontario Regional Committee, Communist Party to Gladstone Murray (CBC), February 25, 1939, National Archives of Canada, RG 41, vol. 260, file 11-40-3.

14. Anne F. MacLennan, "Cultural Imperialism of the North? The Expansion of CBC's Northern Service," *The Radio Journal*, Forthcoming.

15. Canada, Dominion Bureau of Statistics, General Statistics Branch, *The Canada Year Book 1930*, (Ottawa: King's Printer, 1930), 602.

16. "Radio Licence Worries," *The Montréal Gazette*, April 14 1937, National Archives of Canada, RG 12, vol. 557, file 1672-4, vol. 5.

17. "Pseudo-Vendors, Radio Licences Said Canvassing," *Evening Edition The Citizen Saint John*, April 9, 1937, National Archives of Canada, RG 12, vol. 557, file 1672-4, vol. 5.

18. "Are You Listening?" *Toronto Star*, February 16, 1937, National Archives of Canada, RG 12, vol. 557, file 1672-4, vol. 5.

19. CBC Internal Memo, "Complaints received regarding increased radio licence 3 fee," March 19, 1938, National Archives of Canada, RG 41, vol. 394, file 21-17 (pt. 2).

20. "Canadian Broadcasting System — Or Is It?" Newspaper clipping, National Archives of Canada, RG 41, vol. 394, file 21-17 (pt. 1).

21. Ibid.

22. Handwritten minutes of the Steadfast Social Credit Group #901, Falcon, Alberta, National Archives of Canada, RG 41, vol. 394, file 21-17 (pt. 1).

23. Beyond the fees, radios themselves were not always accessible commodities. Prices for radios fluctuated in the 1920s and 1930s. In 1925 the average unit price was $46.95 (Canada, Department of Trade and Commerce, Dominion Bureau of Statistics, Mining, Metallurgical and Chemical Branch, *Manufactures of the Non-Ferrous Metals in Canada 1935 and 1936* (Ottawa: King's Printer, 1939), Table 111, 79). By 1930 the average price of radio was $112.87, keeping in mind that an average male worker's yearly wages in 1931 were $927 (Canada, Department of Trade and Commerce, Dominion Bureau of Statistics, Mining, Metallurgical and Chemical Branch, *Manufactures of the Non-Ferrous Metals in Canada 1930-1932* (Ottawa: King's Printer, 1934), 62; Canada, Dominion Bureau of Statistics, Seventh Census of Canada Volume V (Ottawa: King's Printer, 1935), 2-3. The increase in price can be accounted for by the introduction of larger and more ornate radios in cabinets. In 1931 tabletop radios were introduced so the average price fell to $63.61 (*Manufactures of the Non-Ferrous Metals in Canada 1930-1932* (Ottawa: King's Printer, 1934), 62). Originally, the average listeners and their less expensive radios were accommodated by regulations restricting the allocation of frequencies to broadcasters, but the Canadian listeners themselves were as resistant to regulation as broadcasters. The falling price of radio receivers and the availability of local stations were followed by the elimination of almost all shared frequencies in 1933.

24. Canada, Dominion Bureau of Statistics, General Statistics Branch, The Canada Year Book 1930 (Ottawa: King's Printer, 1930), 602.

25. K. A. MacKinnon, Engineering Division, "A Report on the Interference from Mexican Stations on the 840, 910 and 960 Kc. Channels," December 12, 1935, National Archives of Canada MG 30, A42, vol. 25, file 143.

26. Gene Fowler and Bill Crawford, *Border Radio: Quarks, Yodelers, Pitchmen, Psychics, and Other Amazing Broadcasters of the American Airwaves* (Austin: University of Texas Press, 2002), 22.

27. Ibid, 25-26.

28. Ibid, 26.

29. Ibid, vii-viii.

30. "Sensational Disclosures of Illegal Radio Sending Sets Made in Montreal," *Quebec Chronicle Telegraph*, March 29, National Archives of Canada, RG 12, vol. 557, file 1672-4, vol. 5.

MATTA

CHAPTER 4

Freedom Soundz
*A Programmer's Journey Beyond
Licensed Community Radio*

Sheila Nopper

BACK IN THE LATE 1980S WHEN I BECAME A VOLUNteer programmer at CIUT, one of several campus-based community radio stations in Toronto, I never thought much about the need for pirate radio as a radical alternative to existing stations. Nor had I considered its significance as a political statement in and of itself — at least not here in Canada. Sure, I had been inspired by stories about the subversive use of low-power radio transmitters by the guerilla freedom fighters in the mountains of El Salvador during their struggle to resist neo-colonialism in the 1970s and 1980s. I had also heard about Britain's pirate deejays, some of whom beamed their programs ashore from boats in the English Channel and, by the mid-1990s, I was reading about the blossoming free radio movement throughout the United States.

But the situation in Canada was different — or so I thought at the time. Access to licensed radio here was more readily available, particularly in urban centres where community stations were often associated with, and located on, a university campus. Furthermore, the Canadian Radio-television and Telecommunications Commission (CRTC), which supervises and regulates all broadcasting and telecommunications in the country, explained in a policy document that:

> The primary role of these stations is to provide alternative program-

ming such as music, especially Canadian music, not generally heard on commercial stations ... [and] in-depth spoken word programming targeted to specialized groups within the country[1]

At the behest of representatives from community stations, the CRTC elaborated upon this definition to specify "programming serving the needs of socially, culturally, politically and economically disadvantaged groups within the community."[2]

In effect, these stations were expected to stretch and challenge the boundaries of political discourse and pop-oriented music programming. Consequently, when on air, I freely critiqued what I considered to be the injustices of our society. Also, through the musical selections I chose to play, and the musicians I interviewed who advocated social change, I provided a forum for people who were, at best, under-represented or, at worst, ridiculed or dismissed altogether in mainstream media, to discuss their concerns, analyses and ideas for creating a more egalitarian society. Yet, it was the calibre of the investigative journalism on the station's spoken word programming, particularly the public affairs show, *Caffeine Free* (heard every weekday morning), which established CIUT's reputation as a credible media source for broadcasting the other side of the story. Consequently, as Meg Borthwick implied in CIUT's Spring 1997 celebratory program guide entitled *Ten Years of Revolutions*:

> [CIUT was] in Kahnewake and Kanesatake during the Oka crisis of 1991, one of a select few media organizations allowed by the Six Nations leaders into their community. These people in crisis placed their trust in us, above most others, because we'd spent years building a relationship of confidence, faith and support with Canada's Aboriginal communities.[3]

Given the existing freedom to broadcast contentious political issues from the perspective of those who are not in power, why, I wondered, would an individual or a small group of people choose to assume all the risks associated with operating what was considered to be an illegal station when there was ample freedom to do what I considered at the time to be radical programming on existing licensed community radio stations?

Although seeking the answer to this question was not a quest I embarked on intentionally, in retrospect, it seems that the combination of my anarchist sensibilities and my passion for radio led me on an illuminating journey of discovery as I evolved organically from *feeling* free as a programmer on CIUT to *being* free as a programmer when, over five years ago, I became one of the cofounders and regular programmers of Tree Frog Radio, a pirate station.

My first step on that journey occurred back in the late 1970s when I was listening to feature documentaries on CBC radio (which were broadcast far more frequently back then before a succession of budget cuts severed much of the funding necessary to produce such programming). Suddenly, I would feel compelled to stop whatever else I was doing while listening to the radio and I would plunk myself down in a chair to focus all of my attention on the audio collage of stories woven into an engaging narrative tapestry. It was the raw sound of the voice — the sighs, the breath and the silent spaces in between the words as much as the words themselves — that often revealed the essence of that storyteller's truth. Isolating a voice from the often distorted filter of visual dimensions that tend to (mis)identify a person enhances the likelihood that the layered emotional nuances contained within (and between) those vocal sounds will resonate directly from inside the speaker to inside the listener. The experience of this intimate type of radio documentary can be profound because it enables the listener to more consciously empathize with the storyteller's experience through their common humanity.

It was through those audio soundscapes that I recognized the potential power of radio to be a transformative medium that could open our minds to different perspectives and ways of being in the world. That new awareness could then motivate people to make ourselves, our community and the world a more inclusive and accepting place — one that recognizes and appreciates differences, but also nurtures mutual aid. Though I had not yet considered, nor experienced firsthand, how the power dynamics associated with who controls the medium itself could amplify — or diminish — such transformative possibilities, I started down the path to that destination when I began to dream of one day producing and broadcasting an inspiring radio documentary.

By the late 1970s, I also became aware of different audio vibrations emanating from the more grassroots, laid-back approach of volunteer programmers broadcasting an alternative to the dominant culture from Carleton University's CKCU in Ottawa, which in 1975 became the first campus-based licensed community radio station in the country. Also, when not tuning into radio, I was now listening and dancing almost exclusively to the potent liberating message, and sticky dance rhythms, of roots reggae and dub poetry. Several years later, I moved to Toronto where I could immerse myself in all of these passions.

CIUT 89.5 FM

Unlike for both commercial radio and the CBC, community radio access didn't require a degree in broadcasting or equivalent professional training. It was open to anyone eager to explore and experiment with their creative do-it-yourself (DIY) projects, and it offered people ample opportunities to learn and share the skills pertaining to radio production with each other.

This open-door policy regarding access and artistic expression was appealing to media and cultural activists, like myself, who appreciated the eclectic live programming that was neither catering to the pretensions of professionalism nor silencing critical analyses of the status quo and the corporate state. Unlike commercial stations that dictated market-driven, computer-generated playlists to deejays, community radio programmers were instead free to choose their own songs as long as they complied with the station's promise of performance agreement with the CRTC. Soon after CIUT's application for an FM licence was approved in 1987 (successfully culminating a 20-year-long dream), I became one of about 250 volunteers who, along with several paid staff, operated the station 24 hours a day, seven days a week, to broadcast 15,000 watts of programming (65 percent music and 35 percent spoken word) throughout southern Ontario and upstate New York.

Before becoming a programmer, however, we had to attend an orientation workshop, where a summary of CRTC regulations and CIUT policies were reviewed, and a technical workshop, which provided an overview of, and hands-on experience with, operating the station's broadcast technology. Prior to being considered as a music program host, an applicant also had to demonstrate their competence by producing a live one-hour program in the presence of a music committee member (one of a small group of elected volunteer music programmers).

CIUT established three primary categories of music programs: a one (or two) hour *exploration* show provided an in-depth context and analysis of a particular genre of music, or music of an identifiable group; the two (or three) hour *open format* program featured a diverse range of musical genres; and a *jazz* show could include a wide selection within a particular style or time period. Each category was assigned variable ratios specifying what percentage of songs played were required to have been released within the last month (new releases), or within the last year, as well as the ideal number of total hours on the weekly schedule the station could accommodate for each category. In addition, the CRTC's cultural policy guidelines specified that we were required to broadcast 30 percent (now 35 percent) Canadian Content (CanCon). Such classifications, which have been somewhat modified by the CRTC over the years, were acceptable if two of the following three criteria were present: the song contained music or lyrics that were composed by a Canadian, the instrumentation or lyrics were principally performed by a Canadian, or the entire performance was either performed or recorded in Canada. While, at the time, I appreciated the sentiments of these well-intentioned guidelines — established by the CRTC in an attempt to preserve Canadian culture and prevent it from being washed away by the tidal wave of American culture that floods in daily from south of the 49th parallel — as I will comment upon later in this essay, after my experience with pirate radio broadcasting, I have re-evaluated the necessity for that regulation.

There were also variable spoken word obligations for music shows. For example, *open format* programs were required to include about 10 minutes per hour of commentary whereas *exploration* and *jazz* shows included 15-20 minutes. Shows scheduled between midnight and 6 a.m., however, were not regulated by the CRTC and therefore were allowed more freedom for experimentation, but, as we shall see, this time slot later became an opening wedge for corporate-produced programming. That scenario, however, appears further down the road.

Reggae Riddims

In 1988 we enthusiastically embraced the new possibilities available to us for creating grassroots programming. That was the year one of my Jamaican-Canadian friends and I submitted an application to CIUT's music committee to co-host a music show called *reggae riddims*. The committee subsequently invited us to attend one of their meetings to discuss our proposal. According to its mandate, the committee would then make a recommendation to the program director and, if approved, the applicant would be assigned a slot on the schedule as one became available. Some of the criteria considered included the uniqueness of the program, the overall knowledge of — and personal collection of — the related music by the host(s), the satisfactory completion of technical training and awareness of on-air obligations. Due in part to the overall popularity of reggae music in Toronto, and the substantial size of the Caribbean community in the region, we were promptly given a prime time slot on Saturday night from 10 p.m. to midnight.

Witches Brew

Obtaining approval to produce a music program, however, wasn't always that simple. Years after leaving *reggae riddims* and having worked on various radio projects, in 1994 I decided to write a proposal to produce *witches brew*, an *exploration* show that would feature women's music and would include a feminist critique of related issues because, although there was a spoken word program dealing with feminist issues, there were no music programs that focused on women's music. I also perused playlists and concluded that only about 10 percent of the music played on the station was by women and only 10 percent of the music programmers were women. I thought that the substantially higher percentage of women involved in spoken word programming was due, in part, to the fact that, unlike music programmers, they were not required to operate the technology as well as host their show. The impact of childhood gender conditioning, which discourages females from being musicians as well as to avoid technology, seemed blatantly evident here. I thought it was important to feature women's music to give it a stronger presence, and, hopefully, to encourage other women to consider becoming music programmers and/or musicians.

Though my initial proposal was approved by the music committee, the program director rejected it because, he claimed, women were not an "identifiable group," and because the music I intended to play would cover a broad range of styles, it would actually be a mosaic program. I did some research and presented a revised proposal with a more cogent argument to demonstrate my claim that women were an identifiable group, and also pointed out the double standards of station policy by positing the question: "Why is it that two music shows aired on the station, one featuring 'Canadian' music, another featuring

'African' music, were accepted as representing identifiable groups but women were not?" To his credit, the program director then replied, "You've convinced me," and approved my show.

On-Air Obligations

Though some flexibility was accommodated at CIUT, all programmers were expected to adhere to the directives outlined on the station's "traffic log," prepared by one of the paid staff, which usually meant we had to interrupt our music sets approximately every 15-20 minutes to play a combination of local commercials, program promotions and public service announcements. Upon completion of this task, we had to indicate the time they were broadcast and sign our initials on the traffic log sheet. During some of these music breaks, we would also identify ourselves, the show and the station, as well as comment on the music we had just played or were about to play. We also had to maintain sequential playlists of the songs we broadcast during our show, which included the album or CD from which the song was taken, its musical genre, release date and CanCon status. Furthermore, all programs were monitored annually by committee members and the program director to determine if the hosts were complying with CRTC regulations and station policies. In addition, we were required to produce a 30-60 second promotional spot for our show, which would then be itemized and integrated into the station's traffic log and played periodically by other programmers.

While some profanity was acceptable when used to artistically emphasize a political stance, in order to avoid a potential complaint from listeners, we were encouraged under such circumstances to preface that portion of the broadcast with a disclaimer to warn people that they might find the following song or commentary offensive. In the midst of juggling all this bureaucratic documentation with cueing up our music and discussing the music on air, programmers also had to answer phone calls from listeners who might want more information about a particular artist, or to simply chat with the deejay for a few minutes. While these spontaneous interruptions magnified the chaos of our multi-tasking, the phone calls were often appreciated because they gave us some immediate connection with, and feedback from, our listeners.

Producing a live radio show on CIUT certainly generated an exhilarating adrenaline rush, yet it paled in comparison to the hyper-intense energy output demanded during the frantic frenzy of the annual live, on-air fundraising drive. Along with what was then an annual $5 levy from full-time undergraduate students at the University of Toronto, and money collected through the sale of community-based advertising, these marathons provided the core funding for the station's operating budget. During the fundraising drives, it was necessary to sustain a heightened level of excitement as we interrupted songs to remind people of what it takes to keep such a radio station on air and to lure listeners to pledge a specific amount of money by offering them various products, donated by individuals and small businesses in the community, such as tickets to an upcoming concert or theatre production, CDs, books and various artistic items. The programmers were also encouraged to at least attempt to solicit some of these enticing items ourselves prior to "the drive." Many volunteers, whether program hosts, technicians or administrative staff, also offered support to programmers by joining them on air to hype their show and the station.

While in some ways this collaborative method of acquiring funds helped to unify the volunteers as we worked toward the common goal of keeping the station — and our shows — on air, its downside was that it tended, however inadvertently, to become a popularity contest based on which shows tallied up the most money. For a popular show like *reggae riddims*, the atmosphere felt like a party with listeners frequently calling to offer pledges and express their appreciation for the show. Even though it was exhausting, it was also invigorating and a lot of fun. However, I had only been broadcasting *witches brew* for a month when I was subjected to the somewhat humiliating experience of relentlessly begging for pledges when I had not yet established a regular audience for my show, which was not much fun at all.

Most of the Time the Honey Was Sweet

I still have many cherished memories from my seven-year relationship with CIUT. It was the place that offered me the opportunity to fulfill my dream of producing a radio documentary. With free access to the station's analog recording and editing equipment (we were approaching, but not yet on the verge of the digital revolution), I learned how to cut and splice ¼-inch reel-to-reel tape and subsequently taught myself how to produce a documentary. It was also through this do-it-yourself process — and my passionate determination to present a comprehensive account of the subversive aspects of the musicians I featured in my documentaries — that I made a conscious decision to abandon my perceived need to produce something "more professional" that might be considered acceptable to CBC. Rather than compromise my principles in order to conform to CBC, standards, which, among other things, severely limited the length of a song that could be played, I decided instead to enjoy the freedom I had at community radio to spontaneously flow with my artistic juices, which was, in itself, liberating.

Back then, in the late 1980s and early 1990s, Toronto was considered to be one of the most multicultural cities in the world and CIUT became a microcosm of that diversity — a vortex where people of various subcultures could gather together to support one another in their pursuits. Whether individually, through cross-cultural collaborations or as a catalyst for marginalized communities' participation, we seized the opportunity to acknowledge and validate a wide variety of voices, political concerns and cultural events. Together, the volunteers at CIUT were like a multicultural swarm of bees buzzing in and around the station's hive; an urban island of rebellious yet celebratory countercultural ideas, music, arts and investigative journalism. And most of the time the honey was sweet.

As the current CIUT website declares, from its inception in 1987, "The next two decades read at times like a weather chart, with highs (mostly programming) and lows (mostly administrative) reflecting both the vibrancy and chaos of a growing organization."[4] Some of the administrative lows alluded to in this somewhat simplified declaration were instrumental in motivating the eventual implementation of drastic measures at the station. However, before those final gusts of wind in the fateful storms descended upon the station in full force, I left CIUT — and Canada — to live temporarily in the United States, where I traversed a new trajectory on the radio spectrum.

The Crossroads

While in the States, I was introduced to, and researched in-depth, what had grown into a free radio movement of direct action, with hundreds of low-power stations defying the dictates of unjust laws that denied them access to the public airwaves. Consequently, in 1997 when I interviewed Carol Denney, cofounder of Free Radio Berkeley, for the then forthcoming book, *Seizing The Airwaves*,[5] I began by asking her the question that had eluded me, "Why micropower [pirate] radio and not community radio?" Her initial response, "Gosh, I wish it didn't just break my heart to answer that question," vividly expressed the anguish she, and many others in the region, felt about the July 1996 hostile takeover, evidenced by the removal of programmers at Berkeley's KPFA — the first independent, listener-supported, licensed radio station in North America — followed by the lockout of programmers and a substantial restructuring of its schedule.

Launched in 1949, after World War II, KPFA was initiated by Lew Hill, a pacifist, conscientious objector and anarchist. According to the *San Francisco Bay Guardian*, the station was "founded as a voice of dissent."[6] It encouraged public debate on contemporary political and social issues and artistic endeavours. This unique radio concept was so successful that KPFA eventually grew into a network of five stations known as Pacifica Radio, and sparked the movement toward the creation of campus-based community radio stations throughout the States and Canada. However, over the ensuing decades, KPFA increasingly mirrored the power imbalances that were being challenged by the politically-charged civil rights, feminist and antiwar movements. The resulting instability at the station was compounded by a gradual series of conservative infiltrations onto its once very liberal board of directors. This takeover shifted the objectives of the station's programming toward a new format that would increase its appeal to the well-heeled segment of its audience at the expense of its more radical programming. As Carol Denney lamented:

> KPFA is what we used to call our community radio station... They are still predominantly a listener-supported station, but it's fairly clear from the programs that they've axed recently, and the volunteer programmers that they've not only axed but blacklisted, that they want a kind of middle-of-the-road station. They see that as more lucrative. It's all about demographics now.[7]

Ironically, one of the programs that was axed was called "Freedom is a Constant Struggle." During my interview with the host of that program (also published in *Seizing the Airwaves*), former Black Panther and prison rights activist, Kiilu Nyasha, accused KPFA of being "disingenuous" with her when they invited her to attend meetings to discuss the imminent changes at the station. "My input was solicited and I certainly gave it," she stated with disdain, "but they never gave me a clue that *Freedom* was going to be cancelled, and that they were going to be throwing out practically all the black and radical programmers. So I was pretty pissed off."[8] She subsequently redirected her media activism through the pirate station, San Francisco Liberation Radio, because, as she explained:

> We can encourage people to get actively involved... to protest police brutality... We don't have to worry about language... So, we really are an alternative in so far as encouraging people, especially poor people and immigrants, to defend against these draconian laws that are coming down and these budget cuts that are about to wipe people out. [Pirate radio] can be a revolutionary tool of communication... educating to liberate. That's what I've been about for years now.[9]

Like Nyasha, Carol Denney noted the free nature of pirate radio when she distinguished it from other community-based stations such as KPFA: "We [pirates] see ourselves as being free of a profit orientation and that is what defines our politics. We have no obligation to sell anything. The truth is not always popular and we can tell it like we see it."[10]

These words of prophesy from Nyasha and Denney reverberated in my mind two years later when I heard disturbing reports of a second aggressive purging of programmers at KPFA in July 1999 that involved the riot squad, and yet another lockout and overhaul of its programming. Closer to familiar territory, several months later CIUT in Toronto became the victim of a similar fate.

The upheaval at CIUT can be traced, in part, to the fact that, during its first thirteen years of FM broadcasting, thirteen managers had their turn at CIUT's helm, which both reflected, and contributed to, the instability of the ship as it careened once too often over turbulent financial waters. Yet, in spite of this frequent management turnover, by 1996 CIUT had acquired a $50,000 surplus. Two years later, however, the station was $100,000 in arrears, and was sinking dangerously close to irretrievable insolvency as its debt climbed at the rate of $8,000 per month.[11] In a hastened attempt to save the capsized ship, in November 1998, CIUT's board of directors voted in favour of a controversial policy that condoned the solicitation of corporate advertising. One month later, when it was revealed that the station could no longer pay its liability insurance premiums, the entire board resigned. The months that followed witnessed an intense ideological battle ground over the contested terrain of CIUT's future identity and shifting power structure that ultimately led to what some have called a coup when, on October 1, 1999, the Student Administrative Council (SAC) of the University of Toronto usurped control of the station and locked out the volunteers.

In conjunction with the lockout, the SAC implemented a slew of comprehensive reforms that included the cancellation all the live, locally-originated overnight programming and the selling of that air time to an internet radio company, Virtually Canadian (IcebergMedia.com), which formatted pre-recorded dance music and supplied CIUT with 25 percent of its then desperately needed annual income.[12] The once-lauded current affairs program, *Caffeine Free*, had two-thirds of its programming time slashed, and one of its most outspoken hosts — a previous board member and spoken word committee chair — Bruce Cattle, was among the handful of veteran programmers who were banned from the station without explanation or due process. A few days later in the University of Toronto's newspaper, *The Varsity*, Cattle declared, "SAC has always pushed for more student involvement at CIUT, but apparently this was just a coverup for another agenda, one of corporatization and privatization."[13]

Such excessive measures by the SAC had a chilling effect on the free speech parameters of the remaining volunteers. This did not deter Rebecca Chua, however, who was then the chair of the station's spoken word committee. In the October 12, 1999 issue of *The Varsity*, she boldly challenged the SAC's inference that it was those programmers that they banned who were at fault for CIUT's demise. As she counter-argued:

> Without doubt, the volunteer programmers were not responsible for the fiscal mismanagement of CIUT. Yet, when the alarm was first raised about extravagant cab fares and cell phone charges, the purchase of laptops and computers incompatible with the system already installed at the station, and a copying machine sans paper or service agreement — not to mention irregularities surrounding advertising commissions — no attempt was made to investigate these charges. Instead, those who asked difficult questions were scapegoated, and the answers swept quickly under the carpet.[14]

Clearly the lack of management accountability at the station had, in effect, morphed into a vendetta against programmers. Interestingly, in a move to further consolidate managerial power, both the spoken word and music committees were later eradicated, and with them some of the station's most participatory democratic procedures. As *Toronto Star* columnist, Peter Goddard, succinctly summarized the ensuing outrage by many volunteers, it is "precisely CIUT's drift to an increasingly centralized, professional style of management that has caused as much dissent as any drift toward corporate advertising."[15]

Now, more than ten years after the lockout at CIUT, according to the current schedule posted on their website, it seems that the overnight schedule has resumed its focus on locally-originated programming. However, the CIUT 89.5FM Media Kit posted on their website is clearly designed to lure lucrative corporate advertising contracts (though local businesses are offered a "discount rate"). Furthermore, the heading on its fourth page is "Demographic Profile," which includes statistics categorized by gender, age, income, household size and even mother tongue (verified in the document by the research company, BBM Canada, for the year 2006[16]), thereby reducing the value of the station's audience to mere marketing bait. As the BBM website boasts, "We provide broadcast measurement and consumer behaviour data ... Our membership includes ... major advertising agencies and national advertisers."[17] As if to corroborate the increasingly professionalized direction of the station, in a 2003 *University of Toronto Magazine* article, CIUT's current manager, Brian Burchell, who by 2002 had guided CIUT out of its financial woes,[18] identified the station's more establishment-oriented shift under his management when he later asserted that, "CIUT is not a political party, and it's not an advocate ... It's in the business of making broadcasting." More to the point, he declared, it is "more medium, less message."[19]

Like a recurring nightmare, early in 2008, organizers of yet another coup infiltrated CIUT's sister station at Ryerson Polytechnic University, CKLN, mirroring the dramatic scenarios played out at KPFA and CIUT a decade earlier. According to the Take Back Our Radio Station website that sprung up as one response to this tragic turn of events, a Special Membership Meeting (comprising Ryerson students and anyone who had donated or volunteered three hours of service) had taken place on February 23, 2008, to discuss their concerns about "the troubling move by the board of directors away from the community vision of the station and towards a corporate and commercialist model."[20] Ninety percent of the 150 members who attended the meeting voted to impeach CKLN's board of directors. The meeting and the vote, however, was ignored by management who, instead proceeded to purge the station of what has tallied up to over 50 shows, including such radical programs as *Anti-Psych Radio*, *Limin' in De African Diaspora* and *Radio Cliteracy*. Callously, management informed the volunteers by email, "Your volunteer services at CKLN are no longer needed effective immediately."[21] Some of these programmers, who defiantly returned to the studio to broadcast their shows, were removed from the station by police.[22]

One year later, the station continued to be plagued by disruptions. As two boards vied for control, management changed the locks and access codes, and protests continued, including a civil disobedience action on March 1, 2009 when two programmers were arrested for barricading themselves inside the station's studio. In addition, some programmers have initiated lawsuits claiming mismanagement and wrongful dismissal.[23] Like KPFA and CIUT, one of the core issues at CKLN was what Toronto *Now* journalist, Paul Terefenko, described as "the old concern that the station's eclectic, street-wise and offbeat lefty political mix is being excised to prepare the way for selling air time to corporate advertisers."[24]

Radio is, indeed, a powerful medium and it seems likely that other campus-based community radio stations in this country are eventually going to arrive at a similarly contentious crossroads, particularly given their financial dependency on both the university with which they are affiliated and the revenue amassed through the sale of advertising, both of which influence the content of what we hear — and, increasingly, don't hear — on these stations. Aggressive shakeups are a sobering reminder of the direct correlation between who has the power to make and implement decisions and the level of freedom that programmers can explore.

As for my personal radio journey, in 2002 I took the fork in the road leading me back to Canada (and away from the States) where I settled on a small island in the Salish Sea off the southwest coast of British Columbia.

Destination Tree Frog Radio

It was in mid-January 2005, on a crisp winter eve, one of the coldest nights people could remember in years, when a handful of programmers gathered together to warm up our island airwaves with the launch of Tree Frog Radio. While we each took turns inside our small trailer studio to spin a few tunes and greet any friends and neighbours who might be listening, the rest of us huddled together round a campfire nearby, swaying to the music, while joking and laughing at the outrageous spectacle of ourselves, this eccentric shivering bunch of dedicated audio rebels celebrating the birth of our homemade pirate radio station.

From its inception as an idea for a community project a year earlier, and throughout its five years of broadcasting, Tree Frog Radio (TFR) has been about working collectively to create and sustain what we described as "a non-profit, commercial-free low-power radio station run by volunteers to foster local culture and build community by sharing our diverse personal passions for music, poetry, stories, rants, and current affairs from a local, regional and global perspective."[25] At one of our first meetings back in 2004, we discussed whether or not our unlicensed station should be clandestine or if we should go public with our intentions. Several people asserted that "there's no way to keep a project like this a secret on the island." The basic attitude was that, "If the community doesn't support us, it won't happen, and we're wasting our time, so we might as well be transparent right from the start." Everyone agreed, so that's what we did.

Anyone from the island who has expressed their desire to be on the radio, and even the occasional kindred spirit from across the moat, has been welcomed into the Tree Frog studio. Whether on the airwaves, or in articles in the community newsletters, or on flyers distributed at fundraising events, we inform people that:

> If you want to do a regular weekly show on Tree Frog Radio (or be a substitute programmer or host just one program), simply drop off your proposal in the envelope at the Free Post under "R" with "Radio Program Proposals" written on it. Someone will then contact you to arrange for you to attend one of our regular meetings so you can meet some of the TFR collective members who will orient you to sta-

tion peculiarities, figure out the best time for you to go on the air and arrange for some initial hands-on assistance with the sound equipment.[26]

Getting your show on air is as simple as that.

Because we refuse to be bound by the restrictive rules and regulations of the CRTC, at Tree Frog Radio we are free to present our programs in any way we want, without having to interrupt the flow to play promos or commercials dictated by managers. Treefroggers collectively manage ourselves. All volunteers, whether deejays and/or technicians, are free to participate in our consensus decision-making process about anything related to the programming, maintenance and operation of the station. Furthermore, we are at liberty to explore the terrain of that freedom because we are uncensored. On Tree Frog Radio, each of us can spontaneously choose the duration of our music sets, as well as when, and how long, we will talk.

I also enjoy the freedom to disregard any need to maintain logs of all the songs I play (though sometimes I do by choice for my own records), and which tracks are new releases and/or CanCon. Interestingly, though Tree Frog Radio doesn't recognize the CRTC regulation that declares 35 percent of music broadcast must be CanCon, given the limited parameters of our broadcast range and our mission statement to "foster local culture and build community," we naturally play what would be considered by the CRTC to be a substantial amount of CanCon.

Personally, I've always emphasized the importance of hearing the voices of women on the radio. Yet, while using my own voice to speak out over the airwaves has represented freedom for me in the past, after listening for a while to another Tree Frog programmer who produces a live sound collage and never speaks on the air — he doesn't even want his show listed on our seasonal program schedule — I experienced an illuminating epiphany of the magnitude of freedom that pirate radio offers. When viewing freedom from another angle, I realized, I could also be liberated by exercising my freedom to choose *not* to speak during my show. This would rarely, if ever, be acceptable on community radio.

Nor would it be considered professional to conduct an interview with someone who is not inside a soundproof booth. But in Tree Frog's small studio we don't have that luxury. Sometimes it makes the interview process there a little awkward. Yet, that awkwardness also creates a more intimate atmosphere, like sitting around one end of the kitchen table. We squeeze in together by the mixing board in our warm, cozy studio, and that down-home sound is conveyed over the airwaves — and it feels good.

As to financing, rather than requiring individual programmers to solicit pledges of money during a live on-air fundraising drive, about once a year Treefroggers organize a social event like a dinner or a dance that is held in a public space so we can meet and talk with our listeners. That way, we also avoid the popularity contest based on which show raised the most funds and, instead, keep our focus on promoting what the station represents as a whole. We have never been in debt.

Sure, we risk being identified by government authorities, tracked down, presented with an order to cease and desist broadcasting and, perhaps also have our transmitter and audio equipment confiscated. But as I have demonstrated in this article, radical voices on licensed community-oriented radio stations continue to face what seems to be an escalating threat of being silenced as political priorities shift with each new station management shuffle. Given the options, it seems pretty obvious to me that the benefits associated with cooperatively operating a pirate radio station far exceed anything a licensed station can offer.

While it is certainly easier in a rural setting, such as our island, to find a frequency on the FM bandwidth that doesn't interfere with the broadcast of other stations, in an urban environment it's still possible to create your own pirate radio station (as the free radio movement in the United States has demonstrated), where the signal and audience would, more likely, be based in a neighbourhood niche. And I can attest to the fact that it is an empowering and extremely satisfying project.

I'm no longer broadcasting from the big city of Toronto in the professional studio of the most powerful campus-based community radio station in the country via 1500 watts of power. And that's just fine by me. Even though I'm sure there are still many sweet programs oozing out of the CIUT honey pot, our intimate low-watt pond of croakin' tree froggers is far more meaningful to me these days. This is where I am truly free. We created our own station, on our own terms, free from hierarchical power structures of authoritarian decision-making, free from the repressive confines of capitalism's obsession with objectifying everything — and everyone — into some kind of branded marketable product. Tree Frog has always been, and continues to be, a raw unrefined concoction of passionate programming that nurtures our community, and I feel honoured to be on board this pirate ship.

NOTES

1. CRTC Policy on Campus/Community Stations: Public Notice CRTC 1992-38, Section 2.
2. Review of the CRTC's Regulations and Policies for Radio.
3. Meg Borthwick, "The FM Project," in *Ten Years of Revolutions: CIUT Program Guide* (Toronto, Spring 1997), 13.
4. CIUT website, http://www.ciut.fm/about.php (accessed May 13, 2009).
5. Sheila Nopper, "People Have No Idea How Powerful They Could Be: An Interview with Carol Denney (Free Radio Berkeley)," in *Seizing The Airwaves: A Free Radio Handbook*, ed. Ron Sakolsky and Stephen Dunifer (San Francisco, California: AK Press, 1998).
6. Belinda Griswold, "Radio Static: Battle for KPFA's soul mirrors fight for the future of the media of America's left," *San Francisco Bay Guardian*, 25 December 1996, 17.
7. Sheila Nopper, "People Have No Idea How Powerful They Could Be: An Interview with Carol Denney (Free Radio Berkeley)," in *Seizing The Airwaves: A Free Radio Handbook*, ed. Ron Sakolsky and Stephen Dunifer (San Francisco, California: AK Press, 1998), 157.
8. Sheila Nopper, "We Have to Make Sure That The Voiceless Have A Voice: An Interview with Kiilu Nyasha (San Francisco Liberation Radio)," in *Seizing The Airwaves: A Free Radio Handbook*, ed. Ron Sakolsky and Stephen Dunifer (San Francisco, California: AK Press, 1998), 123.
9. Ibid, 124-125.
10. Sheila Nopper, "People Have No Idea How Powerful They Could Be: An Interview with Carol Denney (Free Radio Berkeley)," in *Seizing The Airwaves: A Free Radio Handbook*, ed. Ron Sakolsky and Stephen Dunifer (San Francisco, California: AK Press, 1998) 157-158.
11. Bruce Livesey, "What's the frequency, Matt?: U of T student council president Matt Lenner is stuck with the job of bringing order to 'dysfunctional' campus radio station," *Eye*, 9 September 1999, 10-11.
12. Graham F. Scott "On The Air: After 20 years of broadcasts, CIUT is still taking chances," *University of Toronto Magazine*, Winter 07, http://www.magazine.utotonto.ca/07winter/CIUT.asp (accessed May 13, 2009).
13. Andrew Loung, "Radio station silenced: SAC takes over, sweeping reforms shock CIUT radio," *The Varsity*, 4 October 1999, 2.
14. Joe Friesen, "Has Ryerson killed its radio stars?" *The Eyeopener Online*, 31 March 2003, http://www.theeyeopener.com/articles/444-Has-Ryerson-killed-its-radio-stars- (accessed September 10, 2009).

15. Peter Goddard, "CIUT flap putting 'free radio' to test," *Toronto Star*, 5 December 1998, K9.
16. http://www.ciut.fm/pdfs/CIUT89MediaKit.pdf, 4 (accessed May 13, 2009).
17. BBM Canada website, http://www.bbm.ca/en/welcome.html and http://www.bbm.ca/en/about_us.html (accessed May 13, 2009).
18. Joe Friesen, "Has Ryerson killed its radio stars?" *The Eyeopener Online*, 31 March 2003, http://www.theeyeopener.com/articles/444-Has-Ryerson-killed-its-radio-stars- (accessed September 10, 2009).
19. Graham F. Scott, "On The Air: After 20 years of broadcasts CIUT is still taking chances," *University of Toronto Magazine*, Winter 2007, http://www.take5.fm/ontheair.pdf (accessed September 10, 2009).
20. http://takebackourradio.blogspot.com and http://toronto.tao.ca/node/9276 (accessed May 13, 2009).
21. Paul Terefenko, "Out of tune at CKLN: Cops are becoming a sad fixture at Ryerson radio station as labour dispute drags on," Now, 27 August 2008, http://www.nowtoronto.com/news/story.cfm?content=164676 (accessed May 13, 2009).
22. Ibid
23. http://takebackourradio.blogspot.com (accessed May 13, 2009).
24. http://nowtoronto.com/news/story.cfm?content=164676 (accessed May 13, 2009).
25. Tree Frog Radio, "Mission Statement," 2004.
26. Tree Frog Radio, "Beaming out to all prospective radio programmers," 2004.

GORD HILL

CHAPTER 5

Secwepemc Radio
Reclamation of Our Common Property

Neskie Manuel

SECWEPEMC RADIO ORIGINALLY BROADCAST ON THE Neskonlith Reserve from July 2005 to June 2007. We did not get a license from the CRTC when starting because of our position that as aboriginal people we did not give up our right to make use of the electromagnetic spectrum to carry on our traditions, language and culture. Operating this radio station is an expression of who we are as a people; it is the modern version of the campfire where people would share stories. Many Secwepemc stories centre around the adventures of Coyote, the best traveller of the land. He is the best traveller of the land for many reasons, but the main reason is a gift that was given to him by the Creator, the gift of innovation. Coyote was warned by the Creator that this was a powerful gift and he must use it for good and to help the people. We are using this radio to decolonize our airspace, our minds and our hearts. We are not pirates, we are Secwepemc.

For our programming we tried to be as diverse as our population. We chose to focus on youth, language and political programming. In our community the youth enjoy hip hop. For a year we had a locally produced hip hop show. An important goal was to include Secwepemc language programming as much as possible. Three hours a day were devoted to the language, whether songs, classes or news. Whenever our phone number was given out over the air, volunteers were encouraged to use the Secwepemc words. When Secwepemc Radio was beginning, friend and fellow activist Gord Hill produced a show on current political events. We also chose to syndicate shows from the licensed campus and community sector. Some of the shows we chose to rebroadcast were *The Bike Show* from Resonance FM, *Native Solidarity News* from CKUT Montreal, *First Voices Radio* from WBAI, and *Deconstructing Dinner* from Kootenay Coop Radio. Our expression of sovereignty was our ability to choose what was going onto the airwaves and exposing what is going on in other Indigenous communities around the world.

During our time on air we had several opportunities to do some live on-location broadcasts. Our first was at the *Decolonizing Indigenous Youth Conference* hosted by the Lakes Secwepemc Sustainable Building Society at Sxeqeltqin (Adam Lake). This was a great opportunity to have youth share their stories on how they are transforming their lives. Another on-air location was at the *Indigenous Food Sovereignty Conference* held at the Enowkin School on the Penticton Indian Reserve. Both events helped spread the word about our project and our wish that more Indigenous people would operate a radio station under the same principles.

As of this writing, Secwepemc Radio is back on the air. The community is ready for some fresh programming. Looking at what we've learned, we hope to make Secwepemc a permanent fixture in our community to continue the decolonization of the airwaves and our lives.

CHARLES MOSTOLLER

Sonny in the studio at Radio Barriere Lake

CHAPTER 6

Awakening the 'Voice of the Forest'
Radio Barriere Lake

Charles Mostoller

AS A COLD WIND WAS BEATING AGAINST THE THIN WALLS of Marylynn Poucachiche's house in the small Algonquin community of Barriere Lake, a group of elders and adults gathered around a small FM radio, laughing and swaying in their chairs to the country music that blared out of the radio's tiny speakers, while a few small children ran recklessly through the cramped living room. After more than six months in the works, Radio Barriere Lake was finally on the air. To the delight of the community elders, a country playlist was on repeat and throughout the snow-covered reserve, houses were filled with people who danced and laughed into the frozen night. For a moment, the music was interrupted, and over the crackling hiss of the speakers, the sound of women laughing wafted out, followed by a few voices speaking in Algonquin. "*Oma nogom ki ne deta naba mitchikinabiko'inik nodaktcigen*" — "This is Radio Barriere Lake, the 'Voice of the Forest,' broadcasting live from the Rapid Lake reserve, *Kitiganik*."

The moment had been a long time coming. The idea to start a radio station on the reserve had been tossed around by the community many years ago, but had been forgotten in the wake of more pressing political concerns. Barriere Lake is one of the poorest native communities in Canada, yet the community has maintained both the Algonquin language and their traditional hunting way of life, centred around moose, beaver, fish, and fowl caught in their traditional territory.

Although located just five hours northwest of Montréal, today it is perhaps the southernmost example of a traditional, sub-Artic hunting society in Canada. But both the community of Barriere Lake's way of life and language are in danger. The youth of the community, who are spending less and less time in the bush and more time watching TV, are losing the richness of the elders' language, and with it, the elders' extensive bush knowledge. Forced by a lack of schools on the reserve, teenagers must leave the community to study in Val-D'or or Maniwaki, local towns each over 150 kilometres away. The older generations worry about the future of their language, especially when considering that Barriere Lake is one of only four communities in Québec where the youth continue to speak fluent Algonquin.

When at a meeting in Montréal some friends who were doing solidarity work with Barriere Lake told me about the community, I immediately thought about trying to help start a radio station there, based on my own experience witnessing the successes of community radio in indigenous communities in southern Mexico. Weeks later, I talked it over with some people at Barriere Lake, who enthusiastically welcomed the idea. After discussing the radio station at a community meeting, they overwhelmingly decided to go forward with the radio project. In a written proposal, stating the goals and motives for starting a radio station, the community clearly hoped the radio would serve to strengthen the Algonquin language among the youth of the community:

> Our mother tongue is alive and well in our community; we practice it frequently and it is used in our assemblies and meetings. Despite this, it is not guaranteed it will be passed onto our children. Our community faces the destruction of our traditional territory, high rates of unemployment and, most recently, the Algonquin language being discouraged in our schools. All of these things threaten the continuation of our language, and thereby threaten our survival as a people. We see the radio as a means to promote the Algonquin language amongst our youth. All our community members will also have another medium to interact with the Algonquin language, which will serve to strengthen everyone's language and contribute to our community's autonomy.

In a tradition that dates back many generations, the community holds a summer gathering every year where a large part of the community camps together in the bush, celebrating the conviviality of the summer for a few days with canoe and rifle competitions, feasts, bingo and perhaps most importantly, square dancing. It was at one of these gatherings, in the summer of 2008, that the idea to build a radio was first discussed with the community. I arrived with Martin — another supporter who does political work with the community — in the early afternoon, after a long drive north from Montréal. That year, the gathering was being held along a narrow beach on the shores of Lac Larouche, once the site of an important crossing point between the eastern and western wings of Barriere Lake's traditional territory. In a wide clearing along the lakeshore, the community had erected a large wooden stage roofed by a mosaic of blue plastic tarps, which would be the site of the square dancing competition, as well as the nightly bingo tournaments. Stretching far along the beach to both sides of the tarped platform, jacked-up pickup trucks and clusters of nylon tents lined the dirt path that ran alongside the crescent shaped shore.

By mid-afternoon, most of the community had gathered on the beach to watch the highly anticipated canoe races, all the more so because the two of us — the *chigoozis*, or white people — would be racing too. After the 10-kilometer race (in which we came in a close second), Acting-Chief Benjamin Nottaway announced to the community over a bullhorn that after supper there would be a meeting to discuss the radio project. For dinner, each family had prepared food to share and below the blue tarps, row after row of steaming pots were lined up on long tables, buffet style. We *chigoozis* were invited to be served first, as visitors are usually shown enormous generosity by the people of Barriere Lake. This is even truer at community feasts, where guests by rule eat first. In front of a long line of people waiting for food, women standing a row behind the steaming pots began to fill my plate, placing a large spoonful of each dish onto the plate until it was almost overflowing. Moose, beaver, sturgeon, walleye and goose leaked their rich juices into the fried bannock bread that topped my plate.

After dinner, I was walking down the darkening dirt path with Norman Matchewan, one of the community's youth spokespeople, and a key player in the radio project, towards one of the elder's camps where the "radio meeting" was to be held. In the faint-blue twilight, we passed groups of people sitting and laughing around small campfires that sent thick columns of smoke into the already star-pocked sky.

"We're having a radio meeting, over at Michel's camp!" Norman called out in Algonquin towards one of the camps. No one seemed interested.

"Maybe," responded an anonymous voice from beside a campfire. At the next camp we had more luck, and few young men joined us. In a few minutes we arrived at Michel's camp, where about 10 people, including Benjamin, had gathered. No elders were present, nor were many youth, most of whom were off putting on nice clothes for the square dance competition that would start later.

"Bad time to have a meeting, I guess," Norman said, laughing.

I started to explain to the group the ways in which a radio station could help the community strengthen its language, and by extension, its autonomy. But it seemed as though those present already understood all that. They wanted to know about the particulars. How much would it cost? Who would run it? I threw out some ideas, while Norman translated into Algonquin when necessary. I described the equipment we would need to buy — an antenna, transmitter, microphones, etc. — suggesting that we would raise the funds in Montréal for the project and likely find some donated equipment. Ideally, I said, the youth of the community would run the station, and programs could include music, community announcements, children's and women's shows, bush knowledge and Elder's stories — all in Algonquin.

"What about radio bingo?" asked one woman. I had not considered the idea, but all present began to chat excitedly when I responded affirmatively. I asked if anyone has any other ideas, but no one spoke up.

"We like the announcements idea," said Norman. "And the radio bingo." Everyone started to laugh.

Benjamin wanted to know how far the signal would carry. When I said that, at least initially, the range would probably be around 10 kilometers — enough to cover the 59-acre reserve and nearby cabins, but nowhere near the community's entire territory, some 17,000 square kilometers — he looked at me disapprovingly.

"It should reach the whole territory," he said, to which I explained how the radio could progress from a small, low-power station into a more established, high-power station over time. After a flurry of other questions from Norman and Benjamin, they seemed content and confident that the radio would serve the community's interests and resolved to propose the radio project at the next community meeting. According to the plan, myself and other non-native "supporters," as we are known to the community, would provide the equipment and training, and the community youth would run the station. Unfortunately, few youth were present at that first radio meeting, and it would be some time before they would begin to actively participate in the project. A few weeks later, in August of 2008, the station was proposed and discussed at a community meeting and the idea was widely supported, especially among the community's elders.

With regards to the legality of establishing a station on the reserve, we decided that the station would have to be pirate. The Canadian Radio-television Telecommunications Commission (CRTC) requires that all radio stations apply for a permit to begin transmitting, including low-power FM stations. However, the permitting process is long, expensive and is designed for stations that have access to large amounts of funding. In some areas of Canada, issues of frequency scarcity have led the CRTC to crack down on pirate stations and make it harder for low-power stations to obtain permits. Frequency scarcity refers to when the full spectrum (or close to it) of usable FM frequencies in an area are occupied, and is often an issue in urban areas where there are many radio stations. Up in Barriere Lake, however, car radios pick up mostly fuzz on the FM dial and one or two weak French-language AM stations. Furthermore, First Nations communities have protections under Section 35 of the Canadian Constitution, which guarantees them aboriginal rights. Aboriginal rights are loosely defined but include rights to land, rights to hunt, religious rights, among many others, although the government contests whether aboriginal rights include the right to self-governance. Over the last two decades many radio stations have been established on reserves throughout Canada, and those that cannot access a CRTC permit have claimed protection under Section 35.

Throughout the following fall, we held a few training sessions in the community, and in Montréal we rounded up funds for equipment. At community training sessions, we discussed the types of programming the community would like to have, worked on creating station IDs — short, prerecorded clips that tell the listener what station they are hearing — and practiced speaking and reading into the microphone. It was at one of these early training sessions that we came up with a name in Algonquin for the radio, which ended up being a much more difficult process than we had imagined. Although there is a word in Algonquin for radio (*nodaktcigen*), it refers to the actual listening device, or the building that might house a station, but not the more abstract concept of a radio station itself. There was also a debate as to whether the station would be named for Barriere Lake, or Rapid Lake. Rapid Lake is the name of the reserve that was established by the Federal government in 1961 for the people of Barriere Lake. In Algonquin it is called *Kitiganik*, which literally means "the place we were planted." Barriere Lake (or *Mitchikinabik*) is the community's former settlement, flooded in the twenties after the creation of the massive Cabonga hydro-electric dam and reservoir. To this day, the community's identity remains rooted in that place, and they call themselves *Mitchikinabikok Inik* — the people of Barriere Lake. Although the radio broadcasts from Rapid Lake, they decided to call it *Mitchikinabiko'inik Nodaktcigen*, "the people of Barriere Lake's radio," or Radio Barriere Lake in English. Norman and Marylynn, who were present at all the training/planning sessions, also came up with the "Voice of the Forest" part. After we settled on the name, we ran it by some elders who gave it their approval.

During the fall, we sought out equipment and held some fundraising events in Montréal, until we had gathered enough necessary items to begin broadcasting. By early winter, we had a small 15-watt FM transmitter, a 100-watt antenna, two old computers newly outfitted with broadcast and editing software, one microphone, a small mixer, and a few pairs of headphones. In mid-December, we were ready to install the radio in the community. Courtney — another supporter from Montréal and someone who worked tirelessly searching for funding and equipment for the radio, as well as organizing all the training sessions — and I planned to visit the community over a weekend to install the equipment. However, this trip had to be postponed after the diesel generators that power the community failed for days on end. Despite having their land flooded for hydro-electric dams, the community remains unconnected to the hydro grid, relying on dozens of huge, noisy diesel generators for power. After a spate of freezing nights, many houses had to turn on extra space heaters to stay warm, overloading the already to-capacity generators. But by the next weekend, the power was back on and Radio Barriere Lake would soon be on the air.

When we arrived in Barriere Lake the following Friday to install the radio station, we found that many people in the community weren't around. Monthly support cheques had arrived, and most people had made the three hour round trip into town to go to the supermarket and pick up the next few weeks' worth of supplies. We had wanted to have lots of people participate in the installation process, but in the

Antenna at Radio Barriere Lake

end only Norman and Angelo, both youths in their 20s, were available to help. Another bad time for a "radio meeting," I thought. In the coming spring, the community planned on building a cabin for the radio station, but for the time being, we installed the radio in the volunteer-run school at the entrance to the community. Inside, the school was in chaos. Desks and chairs were strewn across the floor and coloured paper, markers and trash lined the floor.

The large, south-facing windows in one of the classrooms let in the weak December sun — but also the freezing cold air — and the building's heater had been shut off due to a lack of heating oil. Cold, but determined, we cleared an area and arranged some desks along the window and Courtney started to unpack the equipment with Norman. Angelo and I found an old piece of scaffolding to use as a ladder, and climbed up onto the steep, snow-covered roof of the school. Overlooking the small community, the closely spaced houses appeared buried by wind-swept snowbanks, extending down a few hundred yards and disappearing into the icy lake. After attaching the two-meter tall antenna to some wooden poles, we mounted it onto a few cinderblocks and covered them with snow and water, which froze the poles in place almost immediately. Come spring, the arrangement would need to be modified, but since the station's placement in the school was only temporary, we hurried off the roof to join the others inside.

Courtney and I showed Norman and Angelo how to set everything up, and then they flipped the switch. We set the transmitter to 107.9 FM and Norman started to speak into the mic to test the signal. We had a boom-box set up, and the sound was coming out loud and clear. Norman got in his car and drove around the reserve to see how well the signal carried, telling people to turn on their radios while he was at it. When he returned, he said that the signal was strong throughout the reserve. We put on some country music, and then Norman got on the mic again, speaking in Algonquin to invite the community to come up to the station and to keep their radios tuned in. Soon the classroom filled up with children of all ages, who ran straight to the microphone, giggling and pushing each other out of the way to be in front of it. "Can I say something?" one girl asked. Norman told her what to say, and timidly she stepped to the microphone to repeat it before running away laughing with her friends. Courtney was trying to get some of the other kids to say something, but most were too shy. As quickly as the room had filled up with kids, suddenly it was empty again. Norman put on some hip hop. I went outside and climbed up on the roof to check on the antenna. The sun was setting over the forest, beaming pink and yellow rays through the icy needles of the pine trees. When I climbed down, a black pickup truck pulled up in front of the school to say hello. As I walked towards it, the window came down and I saw Norman's brother Terry.

"Turn on the radio, 107.9!" I shouted.

"Already got it on," he said, as he turned up the volume on his stereo to blast the hip hop Norman had going. "What time are you doing training tomorrow?" he asked.

"Two o'clock," I responded.

"Alright, see you later then." he said, before driving off.

I walked back inside, surprised, to tell Courtney and Norman. "Everyone's already got their radios on," I said. But before they could respond, someone behind me shouted for Norman. It was Luisa, an elder.

She said a few words in Algonquin, before saying sharply, "No more Rap! Put on some Country! We want to dance!" And then she turned away, heading outside to her truck and driving off. Norman translated. Luisa said that lot of people, including lots of elders, were over at her house hanging out, and they were sick of listening to hip hop. So we put on a large selection of country tunes, set the playlist on repeat with some of the prerecorded station IDs mixed in, locked the door to the school and headed over to Marylynn's house.

Cramped into Marylynn's tiny living room, a group of elders and young adults sat around a small table covered in a flower tablecloth, listening to Hank Williams on the radio. Some young children were running around, prompting Marylynn to tell them to go to sleep in the next room. Norman, Courtney and I took a seat with the others and began to joke about our day at the radio.

"So how many people showed up?" asked Marylynn.

"Just me and Angelo!" responded Norman. "More will come to tomorrow's training, I hope."

For a moment, the music interrupted and a station ID came on the air, one that Marylynn had recorded before with some other women. "*Oma nogom ki ne deta naba mitchikinabiko'inik nodaktcigen*" — "This is Radio Barriere Lake, 'the Voice of the Forest', broadcasting live from the Rapid Lake reserve, *Kitiganik*."

As the music resumed, Marylynn started to laugh, and joked in Algonquin to her father Albert, seated across the table from her, who began to laugh as well.

"I'd never heard that before," she said to me, laughing and apparently very flattered. "I hadn't heard my voice on the radio!"

With some prodding from Marylynn, Albert offered to tell a story for the radio. I grabbed the portable recorder and handed it to Norman. Putting on the headphones, he activated the mic and checked the levels. After pressing record, he asked Albert to begin his story. I listened in fascination to the *tigweygan* (drum) story, even though I only understood a scant few words. After fifteen minutes or so, Albert ended the story and Norman packed up the recorder. As I turned the volume back up on the radio, somebody opened a case of beer and passed a few around. For the next few hours, we were trading stories and laughing the night away, all to the backdrop of Radio Barriere Lake. All over the reserve, people were gathered around together doing the same thing. The frozen night got even colder, and ol' Hank wailed through the speakers of dozens of tiny radios, as if asking the moon to slow its course so the party could go on just a little longer.

The next morning, Marylynn's husband Clayton woke me and handed me the telephone. On the other end was Sonny — a youth who was very enthusiastic about the radio but was in jail at the time for having participated in the peaceful blockade of a nearby highway, which the community had staged in November of 2008. The blockade was held to pressure the Federal government to honour the resource co-management agreements it had signed with the community in 1991 but which had yet to be implemented. Sonny was jailed for a violation of his conditions — to not protest — stemming from an earlier action where Sonny and other community youth were arrested for staging a sit-in at Federal MP Lawrence Cannon's Buckingham office.

"I'm so happy to hear about the radio," he told me. "I can't wait to get out of here so I can help out. Maybe I could talk about my experience in here." We chatted briefly about the radio (collect calls from jail are extremely expensive), and I told Sonny everything we had done so far, and he told me his ideas. We arranged to hold another training weekend a few weeks later, when Sonny would be out of jail and then we said our goodbyes.

"Stay strong," I said.

"You know me," he replied, laughing.

A few weeks later, I made the trip back up to Barriere Lake for another weekend of training with Courtney and Tim, another supporter from Montréal, who was visiting the community for the first time. We brought up a new computer and some better headphones, but we still had not raised enough money to buy any portable recorders. This meant that we would be unable to leave a recorder with the community, a key tool for youth to be able to go out and record stories from elders. For the weekend, at least, Tim had brought his own recorder for the youth to use. This time, more than fifteen people showed up for training. We decided to break into groups. Tim and Courtney would work with Sonny, Jamie and the others to learn how to use the broadcast equipment and editing software for recording. I would accompany Angelo and Paul out to record stories with some elders who had expressed interest in speaking on the radio.

Leaving the volunteer school we walked down the bright treeless road that runs along the top of the community, splitting the school and a baseball field from the densely packed dwellings along the lakeshore. From over one of the tall snowbanks that line the road, a pack of fierce looking dogs greeted us outside a small wooden cabin. Angelo told me it was his grandfather Toby's house. I waited outside on the narrow porch with Paul, while Angelo entered only to return moments later. Toby wanted us to come back in a few hours. We decided to head over to find another elder, Eddie, who lived a few houses over. Another pack of dogs welcomed us, but Eddie wasn't home. Angelo and Paul exchanged a few words in Algonquin. "We're trying to think where to go," Angelo told me. "Not many elders left."

I was looking down over the frozen lake towards the tiny islands that dot the horizon, and asked naively where the elders had gone. Angelo did not respond and when I looked back at him for an answer, his firm gaze met mine, as if imploring respect; he raised his hand dutifully, index finger pointing to the sky. It was then that I fully realized the urgency of the radio's mission.

We headed back up to the school, where groups of people were gathered around the computers, passing the headphones back and forth as they learned how to use the programs. Young children ran rampant through the schoolhouse with their coloured paper and crayons, taking advantage of the unmonitored access to the classrooms. Sonny started to interview some of the little girls, who had just come back from playing broom ball in Val-D'or, practicing for an upcoming tournament. This was the first time Sonny had ever conducted an interview, and the girls were shy. But he had an innate, almost uncanny ability to prod them on, quickly and easily getting them to open up. Soon they were talking so much Sonny had a harder time getting them to speak one at a time. Jamie and Sonny started taking turns introducing songs. Then Sonny cued up some of the interviews he recorded earlier, and announced them in Algonquin. Most people headed home to prepare food a little while later, tired after a long and productive day learning the ropes. Those of us who were left followed Sonny and Angelo over to his grandfather Toby's house.

Toby is the keeper of the community's three *wampum* belts — symbolic strings of coloured beads central to the community's oral history. The first *wampum* represents the community and its history, the second an agreement between the Algonquin peoples and the Québec and Federal governments, and the third represents the community's Great Law. In Toby's dark cabin, we crowded around the small wood stove while Toby and Sonny sat down at a table in the corner. Toby's wife, daughter and a few small children were sitting on the only bed, and turned off the TV as their gazes shifted attentively towards Toby. From a high shelf, Toby pulled out a leather bag, removing from it three replicas of the original *wampum* belts. Sonny set Tim's recorder on the table and Toby began speaking. The complete stories would have taken more than four hours. Instead, Toby gave shortened explanations of the *wampum*'s meaning and history, pointing out different symbols on the belt as he went. Sonny looked on, enthralled. Although they are central to the community's identity, Sonny had only heard the *wampum* stories a few times in his life. For the small children listening, it was probably the first.

As Toby concluded the story behind the first belt — the *Three Figure Wampum* — Sonny prodded him with questions. After a moment of reflection, Toby responded, first pointing at the belts and then into the distance, through the thin walls of his cabin, over the lake, towards the old Barriere Lake settlement to the northwest. Over the forest, the sun had almost set, its dull rays barely penetrating the frost and ice covering the cabin's two windows. Toby's quiet but steady voice competed with the cracking birch in the wood stove and the harsh wind that shook the glowing little windows. After awhile, Toby packed up the *wampum* and returned them to the shelf, before walking over to an old radio near the wood stove. He turned it on as we thanked him and said goodbye. As I stepped outside, I could hear Kenny Roger's warbling voice rattle through the speakers.

Outside the thermometer read minus 30 degrees, and Marylynn's house was quickly filling up with people. In the kitchen, Sonny started to tell Clayton and Eddie about the day's training. He appeared ecstatic. Sonny and Norman had taken it upon themselves to take the radio project under their wing, to keep it on air and running smoothly, and planned to start an announcement/news hour and to help others join the project. I told Sonny that in the future it may be possible to have a paid position at the radio, from funds raised through radio bingo, for example. Sonny shook his head disparagingly.

"No," he said bluntly. "If it starts out being volunteer, it should stay volunteer. People should take part because they want to help the community, not to get paid."

I was surprised by this show of selflessness. The community's unemployment rate hovers around 80 percent, so rejecting the idea of being paid for working at the radio was very significant. I realized that most of the people who had gotten involved with the radio thus far shared Sonny's devotion to bettering the situation of his community and maintaining its traditions.

The next day, we were back at the volunteer school with Sonny and a crew of guys, some of whom were getting involved with the radio for the first time. Sonny was working on his interview with Toby from the previous day and showing them how to use the editing software. Paul was on the other computer, lining up songs and downloading music off the internet. His latest selection — a Wu-Tang Clan song — blared through the big black stereo on the other side of the room. With Courtney's help, Sonny and the guys set up an email account for the radio, and a blog to post audio and photos, as well as to link the radio with other aboriginal radio stations.

The sun was setting, and we had to begin the long drive down to *Monyak*, as Montréal is known in Algonquin. We said our goodbyes, promising to return in a few weeks with more equipment, hopefully including some portable recorders. As we left the school, a few more young men showed up, joining the five or six others gathered with Sonny around the mic and computer. We made the rounds of the reserve quickly, stopping in houses and thanking everyone for their hospitality and promising to return soon. *Migwetch* (thank you) they said. *Kanagootsanun*, we replied. You're welcome.

Soon we were in the car driving down the narrow, snow-covered road that leads from the reserve to the highway a few kilometers away. We turned on the radio and set it to 107.9 FM, just in time to catch the end of Sonny introducing a song. "... listening to Radio Barriere Lake, *Mitchikinabiko'inik nodaktcigen*, the 'Voice of the Forest' ... *Migwetch*." Out the window, frozen lakes rolled by, disappearing and appearing again in between the thick stands of snowy pines. As we headed south, out of the dense wilderness, kilometer by kilometer heading deeper into so-called "civilization," where native peoples have been forgotten, Radio Barriere Lake's signal got weaker and weaker and soon was lost amidst the fuzz. But up in Rapid Lake, Sonny had just put Toby's *wampum* story on the air, and three generations of Barriere Lake Algonquins were sitting around tiny radios, listening.

TOMÁS HAYEK

Squatting the Airwaves
Pirate Radio as Anarchy in Action

Ron Sakolsky

> *A society which organizes itself without authority, is always in existence, like a seed beneath the snow, buried under the weight of the state and its bureaucracy, capitalism and its waste, privilege and its injustices, nationalism and its suicidal loyalties, religious differences and their superstitious separatism. Far from being a speculative vision of a future society, it is a description of a mode of human organization, rooted in the experience of everyday life, which operates side by side with, and in spite of, the dominant authoritarian trends of our society.*[1]
>
> COLIN WARD

WHEN COLIN WARD FIRST WROTE *ANARCHY IN ACTION* back in 1973, he included many examples of anarchist social organization in the areas of work, play, education and social welfare. Missing in action was pirate radio. Little is said in Ward's book about communications. One might assume that one of the reasons for this omission is because of the conflation of communications with *mass* communications. The assumption being that because of its massive scale, corporate hierarchy, and/or government bureaucracy, radio was not a suitable topic for tracing embryonic anarchist forms or ruminating on anarchist possibilities. Since the birth of the free radio movement, this assumption has been increasingly called into question, especially in relation to the latest developments in micropower broadcasting technology where the transmitter can be as small in size as a loaf of bread.

Radio Waves

While Ward's book favourably references the British squatters' campaign that originated in the sixties, he could not have predicted that by 1979, just across the English Channel, Vrij (Free) Keizer Radio, named after the huge squatted housing complex in Amsterdam's Keizersgracht, would take to the air, broadcasting mainly squatters' movement and resistance news and music, and going live during the big political demonstrations and street riots of the day. Aside from playing this kind of tactical role in defending housing squats as occupied space, outside of capital or government control, pirate radio itself can be understood as a form of squatting. By using direct action, radio pirates can communally seize the airwaves and liberate them from institutional control.

In fact, from the mid-seventies well into the eighties, an explosion of pirate radio stations could be found plying the European airwaves from the studios of Autonomia's Radio Alice in Italy, Radio Libertaire in France, Radio Dreyeckland in Germany, Radio Skokkeland in Denmark and Radio Air Libre in Belgium. In Spain, where an anarchist revolution had been suppressed by General Franco with the assistance of both Hitler and Stalin, within a year of the hated dictator's death, the first free radio stations would surface, including the decidedly anarchist Radio Libertaria in Valencia. Even from the vantage point of Colin Ward's writing outpost in the UK, Radio Arthur would soon make its appearance. Named after union leader Arthur Scargill, its origins can be traced to the galvanizing radical politics of the British coal miners' strike of 1984. The micropower radio movement in the States was born in the late eighties in Springfield, Illinois with Black Liberation Radio, and then consolidated with the impetus of Free Radio Berkeley in the nineties. Though not all of the pirate stations mentioned above were explicitly anarchist, they typically operated on a daily basis in ways that resonate with the nascent anarchist organizational forms profiled by Colin Ward in his book.

Once the free radio movement began to gather steam in North America, would-be Canadian pirates could get a front row seat on the action and, with the ever wider availability of inexpensive micropower equipment, it was only a matter of time before they too would want to participate directly. A contemporary case in point is Tree Frog Radio in British Columbia. This island-based station, with which I have been involved since its inception, has been squatting the airwaves for over five years. From the start it was to be an anarchist-initiated project that would be open to the community as a whole. Not everyone on the station is an anarchist, and not all anarchist programmers are always doing programming with specifically or exclusively anarchist content, but its origins and current organizational context are deeply informed by anarchy.

Tree Frog Radio

What then are Tree Frog Radio's affinities with anarchism in Ward's "everyday" terms? In essence it is the human scale of the relationships within Tree Frog Radio and with its community that has won it broadly based support and widespread participation. As one programmer has explained the appeal of the station, "Big radio always felt so cold and distant. Tree Frog Radio, like our community hall, recycling centre, free store and farmers market, feels involving." Though illegal, because it has been the embodiment of autonomous island culture, it has engendered community involvement. It has motivated community members to nurture and protect it over the course of its history, which began with an on-island showing of *Free Radio*, a film about the US pirate radio movement of the nineties, after which around 20 community people began to envision starting their own station. Collectively we combined the programming, technical, fundraising and organizational skills needed to launch Tree Frog Radio.

Most of the folks involved did not bat an eyelash in defence of the concept of legality. Though some concern was expressed about the possibility of a government clampdown, legality was not intrinsically linked to possibility. What was illegal, though riskier, was not necessarily dismissed as impossible. Of course, it helped that the island had long been conducive to libertarian living arrangements that were appreciated even by those islanders who would not necessarily identify as squatters or anarchists. In regard to the anti-authoritarian nature of island culture, many of the bohemian residents who came to live here in the seventies were artists, poets, hippies and Vietnam-era draft dodgers. While island demographics have changed over the years, the steady stream of free spirits has never really dried up. Most emblematic of an anarchist trace that is still very much in evidence on-island is the fact that we have no police. Because something so seemingly impossible as living in a place without cops is indeed possible here, islanders are often more receptive than most people to imagining the creation of other autonomous zones. It is precisely this everyday sense of demanding the impossible that animates Tree Frog Radio. With this open attitude in mind, I will now explore the anarchist implications of the station's libertarian organizational structures, such as community participation, volunteer labour, commercial-free programming, grassroots fundraising, consensus decision-making and community self-defence.

As to community participation, the station was started and continues to flourish as a result of the sweat equity of the community members who built and sustain it. Without resorting to such bureaucratic policies as "outreach," "recruitment" or "affirmative action," from the start the station has quite naturally been a magnet for political, economic and cultural diversity. In addition to the "usual suspects" among anarchists and punks, a grassroots assortment of marginalized islanders, drawn over the years from renters, first generation immigrants, Québécois drifters and those culturally disenfranchised because of their youth have readily taken to the airwaves over the years. Though the station welcomes the participation of all islanders as programmers, it has, from the start, been largely the voice of the voiceless. As one programmer has put it, "Tree Frog Radio provides the realization of the voice many of us have to share but cannot express otherwise."

While many of our programmers do not own land, even those that do tend to be unusual — radical libertarians, back-to-the-landers, co-housing land partners, permaculture activists, unruly wage slaves, gender rebels, counterculture mavens, habitués of the underground economy and eccentrics of all stripes. Up until recently, the local Residents Association had been called the Ratepayers Association, reflecting in its previous incarnation the assumption that it was the more established property owners on island who were the rightful community decision makers. Of course, the fact is that renters indirectly pay property taxes as is evidenced by the soaring island rents, which are in part a result of the local property owners' ability to pass on their land taxes to their tenants. Yet, even though the name Ratepayers has now been changed to Residents, the fully-enfranchised islander is still unofficially conceived of as an adult property owner. Consequently it is the voice of the more affluent property owner that is heard most often in public debate at Residents Association meetings, and those with little or no legally taxable income from employment or retirement pensions are rarely part of that debate. Though the latter are not officially excluded, the alienating culture of formal meetings can often seem unappetizing or unwelcoming to those on the fringes, who choose instead, intentionally or in effect, to withhold their consent.

At Tree Frog Radio, there is no such aura of propertied legitimacy or elitist atmosphere of entitlement. Instead the station's freewheeling lack of formalities attracts a different type of participant than the Residents Association. On the airwaves, the voice of the propertyless or atypical property owner holds centre stage. Though the latter might own land, they do not claim a privileged status or act the part of landed gentry. Consequently, the political opinions expressed on our shows offer the listener access to a much broader spectrum of island politics than one can be exposed to by attending a Residents Association meeting, where, even with the best of intentions, the participatory spirit is stifled by the straitjacket of Roberts Rules of Order.

Another group that is represented on the station in ways that they are not elsewhere in the general cultural and political life of the community are recent immigrants. For example, in the entire region, there is no place on the radio dial other than Tree Frog where you can regularly hear local political commentary on island issues, listen to a scathing critique of Canadian domestic repression of indigenous peoples or get no-holds-barred commentary on the government's dirty little war in Afghanistan; all from the "outsider" perspective of a programmer who is a first generation immigrant of Middle Eastern descent. Moreover, it is not unusual to hear a wide variety of music programming by our deejays, with some vocals in Farsi, Czech, Yiddish, Yoruba or Kwakwaka'wakw, just to name a few languages that would never otherwise be heard in the public sphere on island.

Beyond recent immigrants, Québécois culture quickly found a voice on Tree Frog Radio as well. While in Eastern Canada, the politics of the French language is often hotly contested, in British Columbia, far from Québec, there is little in the way of a public voice for Francophone culture. Yet for the first several years, Tree Frog broadcast a weekly program hosted by a woman of Québecois heritage until she returned to Quebec City in 2008, featuring French music and culture, which was presented entirely in that language. In a country that pays lip service to bilingualism, not even the nearest licensed community radio station within listening range provided such a service until much later.

As to island youth, we have had two shows by deejays who are under 18 years of age, one of whom started at age 14 during the early days of the station and another who began his show at the age of 15 at the end of our fourth year on air. There is simply no public forum on island where a young person would regularly be given similar responsibility, along with an opportunity to learn radio skills while freely designing his/her own show just as the adult programmers do, or be able to participate in programmers' meetings as decision makers, or to deejay at station fundraisers. In essence, Tree Frog is a station whose programmers are drawn from the young and the young at heart. As one now deceased programmer had expressed it, "This experience has revived that sense of awe that I had in my youth when it was all new, when so much was out there to be discovered." Our oldest programmer is in his mid-sixties, an age group that faces similar barriers to doing licensed radio, whether on commercial, public or even community stations, as are encountered by youth in relation to the ageism of conventional broadcasting.

At licensed campus/community radio stations, while the programmers are volunteers, management is typically paid. At Tree Frog where there are no managers, it is an all-volunteer affair. There is no paid staff and so it is all a labour of love (though not without a bit of ego thrown into the mix). All in all, we are a non-hierarchical and self-managing bunch. At this point, Tree Frog meetings (which are open to all programmers and technical support folks) are mainly concerned with making consensual decisions about programming schedules, community fundraisers and station maintenance. In the past, more philosophical and sometimes contentious issues such as whether to accept local business sponsorship for individual programs as a way of fundraising or whether to apply for a low-watt (5-watt probationary) licence were passionately debated. Both ideas were rejected as inappropriate and unnecessary after much internal discussion. In terms of becoming licensed, as expected, not only was the anarchist contingent at the station opposed to going legit, but, for other programmers as well, attempting to become a legal station was generally considered to be too expensive, to involve too long a waiting period and to be too bureaucratic a process to pursue.

By now the station flows pretty smoothly on its own steam with only occasional programmer meetings and the use of a Tree Frog email list for information-sharing and troubleshooting. If an islander wants to do a show, we'll find him/her a slot in the schedule, offer some technical training and put them on air as soon as possible. And because we do not have scheduled programming 24 hours a day, 7 days per week, aside from our publicized programming, we allow for sporadic unscheduled broadcasts by any of our deejays or guest deejays during times when none of our regular programmers are slated to do shows. Since there is no commercial advertising on the station, we rely on grassroots fundraising to pay the bills, which now consist of $35 a month for electricity, and incidental costs incurred in maintaining, upgrading and replacing the equipment. The land on which our tiny trailer/studio sits has been donated to us rent-free, and the trailer itself was sold to us at a discounted rate by an islander who supported our efforts. Much of the consumer electronics that constitute our studio equipment have been scavenged (at the island "free store"), picked up cheap at a nearby thrift store, or were donated (mixer, CD players, turntable, mics and tape decks). Other studio technology has been rebuilt (computer) or, like the mixer and turntable, were eventually purchased new after our original ones had died and could not be easily replaced. We even have a second transmitter that was donated to us for live remote broadcasts by the person who built it at a pirate radio workshop in Berkeley, California.

As to our monthly operations costs, they are paid for by the recycling of bottles. The station has its own Tree Frog bin at the island recycling centre, and anyone can support us by simply depositing their beer and wine bottles in our designated repository. Though all of the other bins are for legal community groups, from the theatre group to the land conservancy, no one seems to mind that we are illegal, since it's obvious that we are providing a service to the community and not harming anyone in the process. If someone disapproves, they can just put their bottles elsewhere. Since our bin is always full of bottles, either our usual compliment of 15 to 25 programmers drink heavily, or the community must think we are doing something right.

At first we had to do fundraising to pay for the trailer and the original radio transmission technology (transmitter, antenna, power supply, compressor/limiter) at a total cost of around $1,500, but by now our only fixed cost is electricity that tends to be payable through our recycling dividends, with the occasional fundraiser used to purchase a piece of equipment. These fundraisers have taken the form of dance parties that are deejayed by our programmers or themed sit-down dinner parties where the cooking is done by us. Both take place at the community hall as would be the case for any other island fundraiser. In each case the person who attends these grassroots fundraisers gets to participate in supporting the station while attending a community social event in return for their contribution to Tree Frog. In the ensuing direct interaction, we get to meet our listeners face-to-face, though the latter happens informally all the time at the local recycling centre, general store, bookstore, bakery or café as well. Typically, the station's supporters use fundraising occasions to get an updated copy of the schedule, arrange to go on-air in the future themselves or tell us personally what they enjoy or find problematic about our shows (any complaints go directly to the programmer rather than to the station as a whole). We also get the occasional unsolicited personal check or cash (the latter is preferred since we have no bank account for obvious reasons) at these fundraising events. Yet, in the eyes of the Canadian government, we at Tree Frog are viewed as lawbreakers simply because we want to communicate with our neighbours without a licence.

Because of our illegal status, and our desire to be "underground" but not entirely clandestine (as is evidenced by this article), we are aware that the possibility exists that we might be in danger of being shut down by Industry Canada, which is the enforcement arm of the Canadian Radio–television and Telecommunications Commission (CRTC). However, the CRTC typically operates on a complaint-driven basis, except for when they accidentally come upon a station during their routine survey operations. Therefore, unless someone complains about a station's existence, it is pretty safe. Industry Canada does not have the mandate, budget or staff to go around looking for pirate stations without a prior complaint. Complaints are typically from commercial broadcasters in relation to pirate signals that they contend are interfering with their licensed signal. Therefore, unless a pirate station is intentionally trying to interfere with the CBC or a corporate station's signal (and most are not), the chances of drawing a complaint are relatively small, though the risk is still there.

Another kind of possible complaint might come from unintentional interference with low-power tourist information or emergency broadcast frequencies, and so care must be taken to avoid such problematic overlap. Or, a disgruntled listener who is offended by a station's programming and contacts the CRTC can ask them to shut down the station. In general, such complaints typically are the result of a listener being upset by political content, scatological language, denigrating personal innuendo, or can sometimes just stem from a grudge against one or more of the programmers. Rarely, do they take the form of a moral crusade against lawlessness.

At Tree Frog, we are not trying to intentionally interfere with another station's broadcasts by crowding their frequency partly because that would interfere with ours as well, so complaints in that regard are less likely. Moreover, our visible role on the island means that we have confidence enough in community support to risk a complaint. Any islander who complained to the CRTC about us would be depriving the entire community of a cultural amenity that has become quite well entrenched as part of island life at this point. Consequently, they might think twice about attempting to shut us down. As we say, if you don't like what's on Tree Frog Radio, you can become a programmer yourself, change the channel, shut it off, or just choose not to listen in the first place. In terms of the latter options, we do not lose any advertising revenue based on listenership statistics since there is no advertising. This in turn allows us not to have our programming options restricted by the constraints of marketing research studies and "audience share" data.

However, should Industry Canada for some reason be dispatched to come over to the island to ferret us out, warn us to cease and desist, close us down and/or confiscate our equipment; our first line of community self-defence is the ferry. Sympathetic ferry-goers are our early warning system that trouble might be headed our way in the form of an Industry Canada triangulator van. As it stands, whenever an Industry Canada vehicle is noticed getting on the ferry, we usually get a heads-up call from someone. Similarly, many islanders, though not affiliated with the radio station, let us know that they have our backs when it comes to Industry Canada by alerting us as to when it might be prudent to temporarily go off air while the feds are on-island on other business. For example, when the Industry Canada van is scheduled to be on island to check the volunteer fire department's emergency broadcast signal, we usually find out about it through the grapevine so that we can lay low during their visit. And, of course, the various grassroots lines of defence publicly mentioned in the above paragraph do not include more covert means of obtaining sensitive information about regulatory surveillance or the use of subterfuge tactics to keep Industry Canada guessing about our location.

A Tree Frog in the Berry Patch of Anarchy

Tree Frog Radio is both a refusal and an affirmation. It is a refusal of the demeaning and disempowering passivity of the bureaucratic model of licensed mass communications, and it is an affirmation of an everyday anarchism that is rooted in mutual aid and individual freedom. While the squatted airwaves of pirate radio can be seen as an example of Ward's "seed beneath the snow," we can look to the ubiquitous on-island presence of the blackberry vine as a way of expanding upon that metaphor. Since wild blackberry seeds have a hard seed coat, they can remain dormant even under winter snow. Rather than constantly requiring cultivation during the growing season, the self-propagating nature of blackberries, implies instead the opening up of artificially enclosed space for wildness to flourish. New blackberry bushes can start not only from seeds (which are typically not planted but spread by animal droppings) but from subsurface rhizomes or crown re-growth.

Stephen Collis has expressed the affinity between the humble blackberry and anarchy in his poem, "Blackberries,"[2] which he read here one summer evening in 2007. Here is an excerpt:

> the fruit which I celebrate
> growing everywhere we cannot purchase
>
> what no one owns shared
> thus our blackberries remnant commons

Unlike the garden variety blackberry, which might be compared to licensed radio, the notoriously difficult to control wild blackberry that is capable of springing up anywhere, might be likened to the unruliness of the squatted frequencies of pirate radio. In essence, the gardener's nightmare of a wild blackberry invasion might alternatively be understood as the gatherer's utopian dream of Big Rock Candy Mountain ease and abundance. In fact, the relationship between the gardener and the gatherer are not necessarily mutually exclusive in that the same person might be engaged in both activities. One person's steadfast commitment to gardening a plot of land need not be condemned in order to appreciate the wandering life of the gatherer and vice versa. For some, it is finding the right balance between the two that makes the whole meaningful.

In the case of Tree Frog Radio, it has been the community that has provided the space and the nurturing soil, with the spark of direct action generating enough light and heat to facilitate the initial growth. However, once up and running, like a spreading underground rhizome, the subversive tendrils of free radio can spontaneously proliferate with the brambled tenacity of wild island blackberries.

NOTES

This article is dedicated to all Tree Frog programmers and our ace tech support crew for providing the energy which animates the station, and to our community which has enabled us to flourish. Personal thanks to all Tree Frog participants for their encouragement and support in the writing of this article, and particularly to Bruce, Jerry and Robert respectively for allowing me to quote their words on what the radio station means to them.

1. Colin Ward, *Anarchy in Action*, (London: Freedom Press, 1973/82), 14.
2. Stephen Collis, *Blackberries*, (Toronto: Book Thug, 2005/06), 15 and 35.

GRETCHEN KING

Free Radio Tent City banner

CHAPTER 8

Amplifying Resistance
Pirate Radio as Protest Tactic

Andrea Langlois and Gretchen King

GRIPPED TIGHTLY IN THE HANDS OF ACTIVISTS, A BANner reading "Free Radio Tent City" was marched into Montréal's Lafontaine Park on July 3, 2003, announcing the Radio Taktic pirate station to the world. Or, if not to the world, at least to a park full of activists in the process of pitching tents and preparing to squat a part of the 40-hectare park in Montréal's inner city. Radio Taktic activists sought to bring together voices denouncing the city's housing policies and the plight of the homeless. Unlike the subtext of the protest slogan, "The whole world is watching," the action of taking over the airwaves during the tent city was not intended to launch the action into the sphere of corporate media for the whole world to see. Instead Radio Taktic, at 104.9 on the FM dial, played an important role in the protest — supporting it strategically and amplifying the voices of those typically silenced in our society.[1] Radio Taktic's equipment was used in three broadcasts during political protests that summer. For Montréal, a city with many media activists, the creation of Radio Taktic was embedded in a community committed to using media tools to support the struggle. For these activists, access to the media is a central element of social justice work. Unlicensed radio is intrinsically a contestation over private property and the power concentrated in media institutions; it is about the creation of autonomous zones in which alternative forms of culture can be created and diffused.

Autonomous media,[2] ranging from pamphlets to zines, pirate radio, and websites, have long been essential elements of the activist toolkit. By creating their own media, individuals and groups involved in social justice struggles take the representation of social movements into their own hands. These media become spaces within which symbolic power is contested — the power to communicate diverse realities. In the case of Radio Taktic's broadcast at Tent City, this meant airing the unmediated, live voices of those impacted by homelessness and poverty — voices often marginalized, excluded or stereotyped in the mainstream. The temporary station was set up under a tree in full view of those taking part in the tent city, with the antenna up amongst its branches, in order to encourage participation.

Accessibility was the key to encouraging participation and bypassing mainstream representations of the action (i.e. those within the corporate news), which was seen as a way to communicate directly to neighbours about Tent City.[3] By airing political analysis and amplifying the voices of those affected by the tent city, the activists were hoping to politicize both residents of the tent city and its neighbours. In a climate where media portrayals of activists encourage mainstream audiences to be intimidated by protest actions, this radio broadcast provided a way for the neighbours to learn about the protest from a distance. The goal was to engender support and to encourage dialogue and participation through the use of a call-in number, breaking down the barrier between Tent City residents and those of the neighbourhood, therefore strengthening the action.

As a form of autonomous media, pirate radio can work in realtime to present information and analysis about social issues and to air reports about the status of protest actions as they unfold. When used as a tactic in protest situations, pirate radio pushes the medium of radio beyond the confines of mere representation. Not only is the broadcast used to communicate a particular view or representation of events, it also becomes a manoeuvre or device for accomplishing the task of protest itself — for example, it may be used to communicate the movements of the police, or to keep supporters engaged with the play-by-play of a protest, indicating what areas need support or what direction the protesters are heading. With the help of eye-witness call-ins, police scanners, or a variety of other information collection mechanisms, pirate radio can become a valuable communications tool.

In the case of the July 2003 Tent City in Montréal, it is hard to assess whether the goals of the pirate radio broadcast would have been fully achieved, because the tent city, which was intended to be a week-long action, was shut down before the first night was over. Police hovered around the edge of the park for the entire day — as people set up their tents, cooked meals, held workshops and broadcast illegally across the airwaves. Late in the evening, police backed a large sound truck into the park and announced that city bylaws empowered them to clear the area at midnight. At this point the broadcast equipment was moved to a nearby house and the broadcast continued throughout the night via web-streaming.

The broadcast itself was limited from the beginning — broadcasting from a tree meant the antenna signal reached only the park and not much beyond. The goal of narrowing the gap between neighbours and Tent City residents was difficult to achieve, partially due to the large size of the park and the fact that the signal did not end up reaching beyond it. Yet, some neighbours did come by to see what Tent City was about and to offer support. (Whether they heard the radio broadcast is impossible to know.) The technological challenges faced by the microstation were compounded by the threats of police violence, the risk of having the broadcasting equipment confiscated and the pre-emptive end to the action. At midnight, the police moved in to crush the tents and homes that had been installed by the people experiencing poverty and their allies. They were met with resistance by the protestors, but within a few hours of their initial orders to vacate the park, the police moved further in and the people drew back. Eventually the police took over the whole of the park, but not before several arrests of Tent City residents and injuries on both sides.

Radio Taktic carried this news — reported via cell phone — streaming resistance music and audio from the day's action over the internet into the night. Although the public space of the park was cleared out, Radio Taktic's resistance continued, albeit from another location. It is important to note that while the police were enforcing city bylaws around camping in parks, they never challenged (or perhaps did not notice) the occupation of the airwaves by Radio Taktic. The right to occupy the public property of the park was apparently under negotiation, but the airwaves remained for the taking.

The act of occupying the airwaves has roots in resistance movements around the world. Radio Alice, set up in Italy during the 1970s to support social movements, was used for many community purposes, from the creation of art to informing activists of the movements of the police during street protests. This tradition continues in many European countries, such as in France.[4] Since the late 1990s, pirate radio has been regularly used in North American protest actions and by social justice movements, such as Y2WTKO Radio run by media activists in Seattle, Washington. As documented on line:

> While throngs of protestors liberated the streets of downtown Seattle during the World Trade Organization convention, several small, independent pirate radio cells liberated the airwaves on Seattle's FM dial to report on the protest and rouse the rabble with incendiary rhetoric and riotous mood music.... Y2WTKO, broadcast into Seattle for five days from a tree on the Olympic Peninsula with music, updates on the demonstration, and relayed news programs from shortwave radio, Olympia's KAOS [a licensed community station], and the local television audio frequencies.[5]

Pirate radio also has strong roots in Mexico, with Radio Insurgente[6] operated by the Zapatistas in Chiapas, and was used in 2006 to support a popular uprising in Oaxaca. In this uprising, radio was an integral component of resistance as a mass movement occupied 14 licensed radio stations and a television station, using the media to mobilise people and fight back against state and federal repression. Since that time, indigenous communities across Oaxaca have started pirate radio stations as part of their political and cultural resistance.[7]

The use of pirate radio within protest movements is a unique form of autonomous media in that the very act of transmitting over the airwaves constitutes an illicit, transgressive action, whatever the content. Other forms of autonomous media, such as open publishing websites like Indymedia.org, push forms of communication into more participatory realms, but are not implicitly transgressive in the same way as pirate radio. When a media tool enables the action of activists, it becomes more than a means of dissemination. It becomes a means of disrupting the social order and transgressing the boundaries of the law. Tactics of transgression are radical, pushing the hegemonic system to its limits, demanding change, and thus heightening the intensity and the immediacy of certain issues. Transgression through pirate radio pushes for the creation of a different world, not by seeking new legislation governing radio waves, but through the creation of a different way of communicating.

Those who use pirate radio as a protest tactic push the boundaries of how we participate in discourse, change how we communicate ideas and information, and question the legitimacy of regulating the air waves as a form of private property. Pirate radio is therefore a form of direct action — a refusal to engage in a politics of appeal to governments, preferring instead the crossing of boundaries, interfering with the state's power and challenging the commercialization of communications. Withdrawing consent by purposefully transgressing the state's laws, pirate radio practitioners engage in a politics of imagining — a politics of creation.

As activist and anthropologist David Graeber suggests, this anarchist practice of challenging authority is what is hopeful about the new social movements engaged in criticizing capitalism.[8] Radical activists seek to dismantle what they see as an illegitimate system, a challenge which is evidenced in their choices of tactics — from organizing unsanctioned marches to occupying the airwaves. Just as asking for a permit from the municipal government to protest in the streets would be seen as giving legitimacy to that institution, an unlawful march challenges the notion that the streets can belong to anyone. As protestors march through the streets crying out the slogan, "Whose streets? Our streets!" they verbalize the fact that whatever the issue at hand, the march itself becomes an act of revolt. An act based in a refusal to consent to the rules and regulations of dominant institutions. As Francis Depuis-Déri, a Montréal based activist and academic, describes in his book about the black bloc:

> Direct actions are also conceived as skirmishes that permit those who participate to send a message onto the public stage and to feel stronger, freer, to deviate from passive citizenship, which encourages liberalism, and to become political agents. These skirmishes are as much micro-revolutions through which activists free themselves (at the risk of their bodies), the space (the street), and the time (a few hours) necessary to live, even for a moment, an intense political experience outside of the norms established by the State.[9]

Depuis-Déri's description of direct action within this context echoes definitions of the Temporary Autonomous Zone (TAZ). The TAZ, a concept developed by Hakim Bey,[10] is defined as a place or endeavour where people can engage in activities and ideas as though capitalist ideologies and state legislation does not apply. TAZs present the possibility of revolution through the creation of spaces in which to live out or to propose and develop alternatives. To borrow the words of David

Graeber, "It's one thing to say, 'Another world is possible.' It's another to experience it, however momentarily."[11]

This comparison sheds light on a common theme in pirate radio practice — that of prioritizing communication over dissemination.[12] With pirate radio there is not as much focus on dissemination in the mass media sense, with a fixation on the numbers of people listening and ratings — but instead on the act of communication. The feeling of freedom and autonomy that transpires by participating in a pirate radio broadcast is one that feeds the desire to communicate and to create alternatives. The goal is not to transpose power from one group to another, but to bypass or confront dominant power structures with an alternative model. And it is within this creation of TAZs that the possibilities for pirate radio lie.

Contesting access to public spaces through squats, protests and the occupation of the airwaves is central to insurrectionist resistance. While tactics are often the centre of debate and controversy within social movements — whether to stage a sit in, carry out an act of civil disobedience, destroy property or write petitions — there is less disagreement around the use of media tools in tactical ways to support and report on activism. Perhaps the lack of controversy can be explained by the highly mediated lives of younger generations who have been influenced by new media, do-it-yourself (DIY) culture, participatory alternative news websites and, more recently, social media. It is quite interesting that in Canadian activist circles the question of whether the airwaves should be free for the using is not a problematic one. The larger and more heated debates centre on whether to engage with mainstream corporate media, which are criticized for their (mis)representation of radical social movements.[13] Therefore, while pirate radio is seldom used in Canadian protest movements, it does not seem to be neglected because taking over the airwaves is controversial or rejected by activists, but simply because the technology or skills may not be readily available. Transmitter building workshops held at Montréal's anarchist bookfair, for example, have always attracted many participants, demonstrating that there are numerous people who want to learn how to build and use radio transmitters. However, there is often not enough experience or equipment to go around.

In the summer of 2003, this was not the case. A transmitter was available, there were activists skilled in setting up and running a pirate radio station and there were several protests planned. Free Tent City

Radio ended up being a trial run for Radio Taktic; a testing ground for what was to be a much larger mobilisation. The World Trade Organization's mini-ministerial meetings were going to be held in Montréal in late July 2003, and FM pirates would be ready. Media activists from across Canada converged on Montréal for the five days of protests, helping to set up an Indymedia Centre (which incorporated print, video and online media), participating in established shows on licensed community radio programs and helping to run the Rock the WTO Radio pirate station.

Large protests at events in other Canadian cities, such as the Summit of the Americas in Québec City in 2001, had illustrated the importance of independent media in mobilisations against neoliberal globalization. At those protests, Indymedia websites, video documentation and other forms of autonomous media served to keep activists informed of what was happening within the protests and to deepen their analysis of the issues at hand. These media were, in short, necessary elements in putting representation back into the hands of those involved, and bringing news and information to those not involved with the mobilisation. The intention behind Rock the WTO Radio was to create a soapbox for the average person to respond to the WTO meetings. The broadcast included live phone calls from the streets and audio recorded from throughout the convergence of resistance. This type of round-the-clock coverage would not have been possible to air within the conventional format of community radio stations. In Montréal there are five community radio stations (CIBL, CINQ, CISM, CKUT all on FM, and CJLO on AM, which was a web station in 2003), all of which have very specific mandates to fulfill with regard to the licences issued to them. Thus, it was agreed that those involved would set up a dedicated FM feed to amplify the resistance to the WTO on the streets, as well as to link up with local community stations to provide reports and to make them available for rebroadcast by posting them on Indymedia websites and that of Rock the WTO Radio.

A location for the broadcast studio was found in the storefront of Montréal's alternative bookstore — then known as Le Librairie Alternative, and now called Insoumise (Dissenter) — and an internet broadcast (i.e. streaming) was set up using open source software. The online element was of tactical importance.[14] Anyone hosting a pirate studio has to consider keeping the location and activity secret, and therefore may face difficulties in recruiting the community to participate, especially potential programmers from marginalized groups already threatened by the law. Over the years, web radio has aided pirate broadcasters in their efforts to create loopholes in the criminalization of their activities by distancing the transmitter from the actual radio station. Although a pirate broadcasting studio can face threats from the police and other authorities, a webcasting studio is a legally legitimate space. In the case of Rock the WTO, the studio at the alternative bookstore was used to stream the station over the internet. A transmitter was then set up at a different location, connected to the internet and the signal from the webcast was broadcast onto the FM dial.

The alternative bookstore was the ideal site for the webcasting studio of Radio Taktic. The volunteers of the bookstore supported the mandate of the radio station and were enthusiastic to help with the work, providing access to the store's internet connection and phone line. Radio Taktic also had a passionate group of activists collaborating to bring the station together. Several people helped to build up the studio, others donated or built equipment and still others focused on programming and publicity for the proposed broadcast. The organization of the studio happened quickly. Others gathered the equipment needed to put the internet feed on the FM dial throughout the WTO meetings. The pirates also worked around the clock choosing a location for the transmitter and antenna, gathering equipment for the FM broadcast, building and installing an antenna and testing the signal. Once all was set up, Rock the WTO Radio was on the air 24 hours a day at 104.5 FM.

The studio setup was simple, with instructions carefully taped up in visible areas. The ease of use was geared towards first-time radio programmers and journalists. The studio table was small, holding a tiny mixer, telephone and the equipment needed for running audio through the board (such as a mini-disc player, CD player and computer). One participant constructed microphone stands out of wood. All of this was set up in the storefront window, allowing the broadcast to be visible from the street, and making it accessible to the public.

Before and during the five days of demonstrations, organized by the convergence of groups that called themselves the Popular Mobilisation Against the WTO, the Radio Taktic group held several meetings to discuss programming. The plan was for producers to record presentations at the teach-ins and speak-outs before the WTO meetings, while others took on the role of calling in reports from the streets. A list of cell phone numbers was on hand in the studio for producers to solicit information and people could email news tips to the studio computer. Independent journalists also worked to collaborate on short daily documentaries that would present an audio representation of the days' events. These reports were available to radio stations through internet distribution (on websites such as Indymedia Montréal and CMAQ.net) and also aired on local community radio stations.

Because the street protests were an important element of the mobilisation, with over 500 people taking to the street for two major marches, Rock the WTO Radio served as an important tool in closing the communication gap between protestors at the march and those elsewhere. Several independent journalists called in or emailed reports regularly, plus the phone list at the studio provided a continuous supply of live voices. The phone number was also posted online, generating calls from supporters in other cities many of whom expressed their criticisms of the WTO and shared alternative visions for a world without corporate rule. Some callers were seeking information about friends who might have been arrested. The storefront space also allowed for engagement with passers-by. At one point the microphone was taken out onto the street and people were interviewed about the WTO and the protests. The result was broader community awareness of the reasons that people were protesting.

The demonstrations were a major presence in the city, blocking downtown traffic and ensuring that business was not "as usual." Temporary autonomous zones (TAZs) other than the pirate radio station spontaneously appeared throughout the city. One such TAZ, the "Green Zone," was set up just outside of the alternative bookstore as a place to hold impromptu meetings, connect with friends, or to take time to eat, rest and recuperate. As activists gathered in the empty lot after a snake march, police in riot gear massed nearby. Reports of the police hiding on side streets came in on the studio phone and Radio Taktic activists rushed to tell the crowd. Simultaneously, the cops moved forward and the demonstrators tried to disperse, but many of them were encircled.

The police action was broadcast live on air. No warning was issued prior to the police charging at the hundreds of people gathered in the lot. Mass confusion ensued, with people running north and south on the street. Many protesters were separated from their friends and affinity groups. Some people in the immobilised crowd had cell phones and called in to Rock the WTO Radio from behind police lines. A microphone was set up on the street for independent reporters to air the voices of those outside of the bookstore. Many people expressed outrage at police actions.

As police officers processed the 240 arrested protesters, which took hours, Radio Taktic activists set up speakers outside so that the crowd gathered on the street could hear the updates and calls coming in from those detained by the cops. Programmers also aired many songs that spoke to police brutality, further highlighting the feelings of rage felt by those on both sides of the police line. Several journalists with Radio Taktic approached the police line to garner more information on what was happening to those being arrested. The police said nothing, but those watching the miscarriage of justice had much to say on Rock the WTO Radio. The programmers in the studio also continued to take calls from listeners, who expressed their solidarity with demonstrators and their outrage at the tactics deployed by the police. Although it was pure chance that the police encirclement of the protestors happened right beside the studio, the swiftness of the media activists in using pirate radio as a tool to expose the police brutality was not. The strategies set in place (call-ins, roaming reporters, etc.), were ideally suited for quick adaptation and broadcasting of up-to-the-minute details of the street actions. They served to create cohesion between what was happening within the encirclement and outside of it, as well as a way of documenting the police repression.

The recording of the broadcasts and their availability online enabled the transmissions to be archived for future purposes. The broadcast hosted by Rock the WTO Radio was carried by web-casters in the United States and Australia. The audio was also rebroadcast by CKLN, a community radio station broadcasting at 88.1 FM in Toronto and archived online, where numerous radio stations downloaded the audio for local rebroadcast. It served to document the events as they occurred, as well as helping to mitigate police repression. In this way, Rock the WTO Radio had an impact on what happened during the protests and it is very likely that the protests would have been less coordinated if the station had not been there to report minute-to-minute news and to inform activists and their allies on what was going on in the street. Furthermore, the station served as a training zone for many activists who did not have any previous experience in broadcasting and who went on to continue their participation in media activism. There is nothing like participation in a relevant, high activity pirate protest to engender a passion for the potential of radio. Lastly, Radio Taktic literally rocked the WTO with countless in-depth discussions and information about the negative consequences of neoliberal globalization. At the end of the protests, some of those involved in the project were so inspired that they planned to keep the station running into the future. These aspirations were not fully realized, but Radio Taktic continued throughout that summer.

The last major mobilisation for Radio Taktic was a live broadcast from a simulated Palestinian refugee camp set up in the empty lot next to the alternative bookstore. This creative display was made up of several tents. One displayed profiles of two refugee brothers — one accepted by Canada, the other rejected — a second displayed audio/video productions of the stories of Palestinian refugees, including interviews with Palestinian refugees living in Montréal who were threatened with deportation back to camps, and another had a map of Palestine. The radio programming highlighted the goals of protestors who set up the interactive display featuring life in refugee tents. The emphasis of Radio Taktic during this symbolic protest was to focus on content about Palestinian refugees, encourage listeners to come to the "camp," and interview the organizers on site.

Although Radio Taktic was a temporary pirate radio project, the lessons it has to teach activists are many. Free radio can be used to disseminate information about protest actions to a whole community, be a part of creative resistance, and, simultaneously, serve the needs of those participating in the protests. Radio Taktic's reach was impressive, not only broadcasting locally, but also bringing the news of the mobilisation to other cities in Canada and other parts of the world via the internet. The decision to use web streaming during the anti-WTO protests is an important lesson of how to occupy the airwaves while maintaining a safe space and an accessible studio. Even if the station was not targeted by police in Montréal, taking steps to ensure that activists would not be directly linked to the illegal action of occupying the airwaves was an important precaution.

In other cities where there have been mass protests against the WTO, Republican National Convention or the Group of Eight (G8), such as in Genoa, London or New York, media centres supporting the actions have been targeted by police. In the case of Radio Alice, operating in the late 1970s in Italy, the station's actions to support street protests resulted in it being shut down by police, and some activists involved in the station were charged with inciting a riot. The station, nevertheless, kept popping up in new locations.[15] In light of the repression that unlicensed stations sometimes experience, the fact that Radio Taktic's occupation of the airwaves was left uncontested by police is significant, especially considering the police repression faced by street protestors. It was, it seemed, literally "off the radar." In Canada most pirate stations do not face legal problems, unless there is a complaint made to the Canadian Radio-television Telecommunications Commission (CRTC), the federal government regulatory body that oversees telecommunication carriers. The threat of legal measures can, nonetheless, be seen as a barrier to people's involvement in setting up pirate radio stations. This is one aspect of pirate radio that sets it apart from other media tools and strategies.

Despite this difference, pirate radio is similar to other media in that it both presents barriers to participation and removes them. Although a small transmitter can be purchased cheaply through the internet and/or built with minimal tools, this can present a barrier to people without funds or the skills needed to solder the transmitter together or set up the antenna and studio. While the skill of hosting radio programming can be learnt and although pirate radio broadcasts have no set rules, other than those agreed upon by those collectively determining the project, being on air does require a certain amount of confidence and skill. Having more experienced radio broadcasters mentor people interested in hosting and deejaying is a great way to open spaces for involvement. Running the control panels in the studio equally requires skills that can be shared. Increasingly, activists with these aptitudes are creating spaces within which to share them and community radio stations are admittedly the training ground for many pirate radio broadcasters. Yet, the benefit of learning within pirate studios is that there is often no lengthy application or volunteer process, as with community radio (see Chapter 4 for more on this). And, because pirate stations do not have to be licensed, they are free from the guidelines that govern licensed stations. Furthermore, pirate radio enables media activists to bring the medium into the streets, thereby meeting the community where they are at.

A later use of pirate radio in Montréal, known as *Sonique Resistance* (Sonic Resistance), was an interesting example of bringing radio to the streets. Sonique Resistance was an effort to build a portable sound system from recycled speaker parts. Activists constructed two homemade speaker boxes that rolled on plastic wheels so that they could be pulled by bicycles. The plan was to use a 1-watt transmitter to broadcast the sound to anyone carrying a boom box, thus amplifying the sound over a larger area. Unfortunately tests of the system were unsuccessful because the commercial radio stations' broadcasts overwhelmed the transmitter. Yet, if Sonique Resistance had been able to afford a higher watt transmitter — which would have required a $100 instead of $20 — their mobile sound system would have had the potential to bring the sounds of resistance to the streets, broadcasting from within protests and marches.

With the combination of technical and volunteer resources, pirate radio can effectively amplify protest action. In activist movements where mutual aid and the sharing of skills, knowledge and resources prevail, pirate radio is an excellent way to engage people in the creation of media content. Bringing a multitude of voices together creates a culture of knowledge in which those affected by oppression and marginalization are valued, abolishing the hierarchy of power so often reproduced in corporate media. By creating new spaces for freedom and autonomy, while also supporting resistance and dissent, pirate radio represents a tactic well suited to activist movements. This chapter outlines just a few examples of the use of pirate radio stations in protest — the future holds many more possibilities.

Shortly before this book went to press, activists converged on Vancouver to protest against the 2010 Olympic Games and the related issues of the criminalization of poverty and Olympic development on unceded indigenous land. Media activists in particular set up a few independent media projects. One was the Vancouver Media Coop,[16] which became the hub of video activism and also had a web radio presence. The second was an artist-run project called "VIVO 2010: Safe Assembly," which included an unlicenced radio transmission with the "goal of facilitating cultural expressions that arise from the community in a lineage of solidarity."[17] The latter station was shut down within 24 hours by Industry Canada officers (wearing Olympic-branded garb) who threatened VIVO as an organization with a fine of $25,000 a day and with fines of $5,000 per day for each individual involved in the broadcasts. The station went off air, yet continued to stream over the internet. Although internet radio is increasingly becoming a force in autonomous media around protests, the Vancouver convergence illustrates that activists, and particularly artists in this case, continue to explore the possible uses of pirate stations in the context of direct action.

As this recent example illustrates, the tactical use of pirate radio no doubt maintains its place in the medley of tools used to amplify resistance and dissent, and, in the words of Felix Guatarri, to create "new space of freedom, self-management, and the fulfillment of the singularities of desire."[8]

NOTES

1. To see a video of the July 2003 Tent City, visit http://vodpod.com/watch/1138340-100-riot-police-evict-tent-city-Montréal-2003?pod=feralvision

2. Autonomous media are a particular kind of alternative media and are directly linked to social movements. They are defined by their openness — in terms of content and membership — and their objective of amplifying the voices of people and groups that normally do not have access to the media. Autonomous media are intended to provide people and communities with information that is alternative to that within the corporate mass media, and audiences are encouraged to participate directly in the production of content. Andrea Langlois and Frederic Dubois, *Autonomous Media: Activating Resistance & Dissent* (Montréal, Canada: Cumulus Press, 2005), http://www.cumuluspress.com/autonomousmedia.html (accessed September 11, 2009).

3. Park Lafontaine, located in the Plateau-Montréal neighbourhood, is over 40 hectares in size and was chosen because of its location and amenities, such as bathrooms.

4. Radio Libertaire has been operating since 1981, http://federation-anarchiste.org/rl/index.html (accessed May 10, 2009).

5. Miskreant, "Aural Assault! Free Radio vs. Free Trade in the Battle of Seattle," www.efn.org/~radio985/AuralA.html (accessed September 11, 2009).

6. See www.radioinsurgente.org (accessed May 10, 2009). According to their website, Radio Insurgente is a FM project which transmits from various places in Chiapas directed to the Zapatista bases, the insurgents and activists, commanders and local people in general. This program is broadcast not only in Spanish, but also in indigenous languages. The program mixes local, national and international news with music, educational and political messages, short stories and radio-novels. Radio Insurgente the media through which the Zapatista communities spread their own music, words and thoughts.

7. Charles Mostoller, "Oaxaca's Media Wars," *Znet*, http://www.zmag.org/znet/viewArticle/17490 (accessed May 10, 2009). Also the film *Un Poquito de Tanta Verdad* (A Little Bit of So Much Truth), produced and directed by Jill Irene Freidberg, Corrugated Films 2008. See also Chapter 2 in this volume.

8. David Graeber, "The New Anarchists," *The New Left Review* 13 (2002), 61-73.

9. My translation. Francis Depuis-Déri, *Les black-blocs: La liberté et l'égalité se manifestent*, (Montréal, Québec: Lux 2003), 39.

10. Temporary Autonomous Zones are described in more detail in Chapter 9. See also the original text: Hakim Bey, *The Temporary Autonomous Zone: Ontological Anarchy, Poetic Terrorism* (New York: Autonomedia, 1991).

11. David Graeber, "The New Anarchists," *The New Left Review* 13 (2002), 70.

12. Anna Friz, Chapter 13 in this volume.

13. For more on the challenges activists face with the mainstream media, see: Andrea M. Langlois, *Mediating Transgressions: The Global Justice Movement and Canadian News Media* (Unpublished Master's thesis: Concordia University, 2004).

14. Web streaming is becoming more common in protest situations. For the September 2009 protests against the G20 summit in Pittsburgh, USA, media activists were planning a "G-Infinity" media centre. The plans for G-Infinity included a strong web radio component. http://pittsburgh.indymedia.org/news/2009/08/31184.php (accessed September 11, 2009).

15. The transcripts of a Radio Alice broadcast from 1977 when leftist and rightist student clashed in an angry but non-violent confrontation on the university campus in Bologna, which turned into a several day battle with police, are available online as part of an article printed in the Toronto publication *Red Menace* in 1978: www.connexions.org/RedMenace/Docs/RM3-RadioAlice.htm.

16. http://www.vancouver.mediacoop.ca.

17. http://videoinstudios.com.

18. Félix Guattari, *Soft Subversions*, ed. Sylvere Lotringer (New York: Semiotext(e), 1996), 73.

SANDRA MORLACCI

VICTORIA'S 3RD PIRATE RADIO MUSIC FESTIVAL

Temporary Autonomous Radio presents:

TAR 99.1 FM

Live On-Air:
Jennifer Louise Taylor
The Pine Family
Jay McLaughlin
Tres Rythm
Gail J Harris
Gerald Fitzella
Wes Borg
Joey Outlaw
The Remanes
And more...

TUNE-IN Thursday, May 7th 2009 7pm to 1am-ish

temporaryautonomousradio@yahoo.com
http://www.myspace.com/temporaryautonomousradio

CHAPTER 9

The Care and Feeding of Temporary Autonomous Radio

Marian van der Zon

A LARGE MAN ENTERED THE BAR WITH A GHETTO BLASter on his shoulder. It was tuned to TAR 99.1 FM, and punk rock music from The Remanes, who were playing live on stage, spewed out. He was already grooving, but as he turned the corner and caught view of the stage, a huge grin covered his face. He yelled his support, which in turn was picked up through the vocal microphones, only to be narrowcast back through his radio and into his ear. He proceeded to rock out on the edge of the mosh pit.

He wasn't the first to wander into the bar that day, called forth by the sounds emanating from the radios across Victoria tuned into TAR 99.1 FM. This was Temporary Autonomous Radio's most recent festival, cheekily being broadcast out of a bar full of people who had been enticed by the live folk, traditional, roots, alt country, rock and punk music being played by the bands taking over the airwaves. Between bands, live interviews were taking place from the corner of the bar — our on-air studio — interspersed with spoken word, sound art and more tunes. Performers had been mentioning the bar name (omitted here so that the venue can be potentially used again) all evening, much to the manager's chagrin, and people were listening in all over the city due to the posters, internet and email lists, word of mouth and even promotion on CFUV, the licensed campus and community station.

Once again, we were taking over the airwaves with pirate radio that was diverse, accessible and free!

TAR began in 2003 as a 2-watt FM radio station that covered a range of about three blocks (when the battery power and weather were good) in Montréal. I had ordered a kit online[1] for $20, and spent a couple of days soldering components onto the circuit board and tuning the coils in order to get onto the FM band. Tuning was a two-person job because the antenna (scavenged from an old television) needed to be positioned in the optimal way, and somebody needed to listen to a radio. The radio needed to be close enough to hear what was happening beside the transmitter, but not so close that it created interference. My friend, who later became known as Pirate Emma P, came to the rescue, helping me out while I bounded from one end of the apartment to the other with the antenna, trying to find the optimal placement on either the front or back balcony. We started to hear faint strains of music coming through the FM dial, fine-tuned a little more, and voila! TAR was on-air for the first time.

I've since grown TAR into a 12-watt[2] station, having caught the bug for unlicensed radio, and wanting a solid signal that would consistently cover a city. The stability of my new transmitter has allowed me to put my time into the organization of the station, the events themselves and all of the elements around it. These include promotion, bringing new folks onboard, mixing, hosting, networking, setting-up, striking (dismantling the equipment) and providing a consistently solid narrowcast, assuming that the antenna is set up correctly. (I use the term narrowcast because we rarely reach beyond a city in terms of range.) But I'm getting ahead of myself.

I was inspired to name the station Temporary Autonomous Radio because I was reading Hakim Bey at the time and delving into his concept of temporary autonomous zones (TAZ).[3] Bey defines TAZs in three primary ways. First, they are freely chosen. Rather than a family based on genetic membership, a TAZ includes a band of individuals, or an intentional affinity group. Second, a TAZ involves the element of festival. Fun and celebration are valued. More than this, a festival cannot happen everyday so it creates special meaning. It is an intense moment, a shift in consciousness. Finally, Bey borrows the term "psychic nomadism" from Gilles Deleuze and Félix Guattari to speak about a state of mind and being. It includes travelers who are curious and adventurous and who are not tied down. The element of psychic nomadism involves intention and a shift from passivity to activity — for TAR it was a perfect fit. It could be a tactic, a place or a platform to speak through airwaves that are ours for the taking, in alternative and temporary ways.

When Bey was developing the concept of temporary autonomous zones, he researched and theorized the "pirate utopias" of the 18th century. These pirate utopias were "intentional communities," places where pirates could live outside the law. This model appealed to me on many levels. While negative connotations exist with regards to the depravity of pirates, there is also much to be gained from an alternate interpretation of these roguish seafarers. They represent freedom and heroism, and so are the role models that I certainly need. In her book, *Bold in Her Breeches: Women Pirates Across the Ages*, Jo Stanley makes the point that "piracy is often in the eye of the beholder."[4] Those in power may pillage and plunder from humans and the environment without consequence. Unlike pirates however, they have protected themselves by laws and institutions of their own making. Noam Chomsky argues that the way pirates are treated — both historically and currently — is political and partisan, and is based on creating meaning around behaviour that is seen to be criminal in one context (in the case of pirates) and legitimized in another (for those who hold the reins of institutionalized power). This holds true for pirate radio as well. The airwaves are considered to be "public" by Canadian regulatory bodies. However, in order to legally access the airwaves, there are numerous barriers — obstacles that prevent the layperson from gaining access, but that favour corporate entities. Because Temporary Autonomous Radio is unlicensed, it has continued to resurface, *uprising* outside of the Canadian nation state's radar in numerous forms in order to remain malleable, in an intentional state of chaos. Ultimately, these elements have facilitated its existence and it continues to grow.

Baby TAR

The original TAR 2-watt FM radio transmitter was about half the size of a business card. I was excited by this small size at first, convinced that mobility would be excellent. While this holds true, the microphone was so sensitive that if I spoke louder than a whisper from five feet away, my levels peaked and threw us into the realm of distortion. While this held lots of potential for sound art, it did not facilitate clarity. Consequently, anyone who spoke loudly would have to be set up outside of the room where the transmitter (with built-in microphone) was located to keep the volume down. In terms of power to run the transmitter, the website where I had purchased the kit advised me to use a 9-volt battery, but when my transmission range began to dwindle after 15 minutes, I needed an alternative solution. I then discovered that I could hook up two 6-volt lantern batteries in series and narrowcast for over five hours without any problems.

I had always wanted the station to be easily accessible and to have an element of spontaneity, so I took an old white lace tablecloth and glued some large black TAR letters onto it. In order to have folks walk in off the street or hear our narrowcast and come looking for us, this makeshift banner was thrown over my Montréal balcony to identify our location and lure potential narrowcasters up my winding rickety staircase. Despite attempts to draw in strangers and neighbours in the early days, it was mostly friends who came by to access the airwaves. People would hang out in the living room listening to radios, while others would take turns on the microphone; telling stories, ranting, playing music, attempting radio karaoke and creating sound. Folks who had never been on-air or used any form of media ventured onto the airwaves for TAR narrowcasts, although they often joined with some level of trepidation. Yet the scale of TAR was so small that it was hard to be completely intimidated and even those who were originally nervous soon overcame it. They became caught up in the convivial atmosphere and the desire to send their messages out to whoever might be listening in their cars, in their homes, on the street or in the living room. For many, participating in TAR entailed a process of building confidence and a way in to other forms of activist media. Some of the folks who were most nervous were women in their 40s and older. In one instance, I had a friend bring his mother, aunt and grandmother. Initially, they stated that they didn't have anything to say. Once they began to tell stories however, and to hear the laughter echo back from the living room, they became comfortable, even ribald. The response to their shared stories and the adventure of participating in pirate radio itself provided empowerment.

TAR Begins to Grow

After moving to the west coast of British Columbia, I continued to use the 2-watt radio transmitter for pirate parties out of my home. We used the transmitter as a sound system, and a friend, Fancy Jenny Fortune, made a Jolly Roger banner for one pirate party. This one far surpassed the former, featuring a white skull with piercing red eyes over two cutlasses. It is the flag that we have continued to fly at every TAR festival since.

Eventually, I tired of the limitations of my 2-watt transmitter, and upgraded to a 12-watt one about the size of a toaster. I ordered it via the internet and it was shipped to my home address from the UK. Though I wondered it if would actually arrive, it passed through customs with an extra duty charge and no more. With this new variable 12-watt transmitter there was still mobility, but there was also the ability to expand TAR. I now had a mobile pirate station that could be used in workshops or for community events, like the pirate radio music festivals that later occurred in Victoria. I was still primarily interested in using TAR to build community and provide easy access to the airwaves, especially for those that don't have such access in our society. Community is built through TAR every time it takes to the airwaves. Individuals who participate find a commonality with one another through the direct action of pirate radio. As well, since it is temporary, there is greater freedom in publicizing events.

Stoked about my new transmitter, I put out the idea of hosting a pirate radio music festival. A number of bands expressed interest, including my own. A few more folks said they would help out with the technical requirements, and, importantly, a friend offered up her funky spa on the top floor of an old heritage house. The space was beautiful, with roofline angles, stained glass and dark wood details. It had numerous rooms, so we could use one for the transmitter (keeping all interference down between the on-air mixing board and transmitter), another for the on-air booth, and the main area for a live band room. Ultimately, the antenna was mounted to the top of an extended microphone stand, out on the back porch, reaching up towards the top of the roof.

The first pirate radio music festival in Victoria occurred in May 2007, for a span of seven hours. We had eight local bands (representing the musical genres of roots, folk, traditional, jazz, punk rock, ska/funk and electronica) perform live for about 30-minutes each. We also had a number of interviews between musical sets on homelessness and local politics. Individuals stepped up to perform rants, spoken word and poetry. The Victoria Anarchist Reading Circle came in for a live discussion, and the MediaNet Soundscape Group presented their recently completed sound art pieces on air.

I had set up a MySpace site for the station[5] to publicize the event by providing the line-up, pictures and information, and this website still exists at the time of this writing. We had posters and handbills printed up and distributed across the city. The location was not disclosed — only those directly involved, who needed to know, were privy to this information. Word-of-mouth and extensive email lists were used to disseminate information about the event. The local licensed campus and community radio station, CFUV, announced the festival over their airwaves. Even CBC Radio 1 caught wind of it and announced it on the Victoria morning news program.

Suffice it to say, all of this made me a little nervous. I knew that in Canada, low-powered radio can continue relatively undisturbed unless it is met with complaints. I knew that after a complaint is lodged, the CRTC (Canadian Radio-television and Telecommunications Commission), through Industry Canada employees, is then compelled to step in and, initially, provide a warning to quit the airwaves. If this is not done, equipment can be confiscated and fines can be levied. If one has positive public support or disinterest, and no complaints are filed, it is possible to continue narrowcasting indefinitely.

Armed with this knowledge, it seemed unlikely Industry Canada employees would find us within the seven hours that our temporary pirate radio music festival took to the air. Nevertheless, the wide publicity through posters across the city and shout-outs over CFUV and CBC made the butterflies swarm in my belly. There was no question about going ahead with the event. The point of advertising openly was to be as inclusive as possible in terms of participants and to reach a wider listenership. Besides, publicizing the event was a political move — after all, the airwaves are ours, we are the public. Still, I eyed up the transmitter, measured the distance to the back door, and was ready to run with it at the slightest provocation. From my perspective, if there was no transmitter, there was no illegal broadcast. Furthermore, the transmitter cost me about $800 and this was a large sum of money for me. Having it confiscated would shut TAR down for a significant period of time before I was able to raise funds for a new transmitter. I even asked three or four friends to keep an eye open for Industry Canada employees, and if they were spotted, to give the word to fly. No one, to my knowledge, complained. No one came to shut us down, or has since.

Regardless, each festival the butterflies are there, sometimes because we up the ante, making it more public, and sometimes because I find out more information. I recently read in the Radiocommunication Act, for example, that you can get a fine of up to $25,000 or a year in jail for an unlicensed broadcast. I haven't heard of a single case of this happening (and have worked hard to find one), but it does give me the jitters. Because we are doing live music, there is a lot more gear involved and it is borrowed from many friends and allies. I've always been a little concerned that other folks' equipment might be confiscated and I'd be on the hook for replacing it.

There are other ways around the law and perhaps I'll explore them in the future. For example, live music could be streamed over the internet in real time.[6] This audio stream could then be picked up at an alternate location and broadcast — or narrowcast — over the city via the FM band. Streaming audio over the internet is legal, so any musical gear affiliated with the live music is then protected against seizure because the transmitter is in a separate location. Until now, however, a few others and I have simply been ready to grab the transmitter and run. Narrowcasting over the FM band speaks directly to the issue of access because many folks can access a radio far more easily then they can access an internet connection to stream a radio station. Those on the lower end of the economic scale are more likely to be able to participate or tune in using a transistor radio, even for individuals who are homeless. It is ironic that we have had listeners tell us about the difficulties they have had tuning in, only to discover that it is because they are linked to radio through cable or satellite. They need only to dumb down their technology in order to access TAR's narrowcast.

TAR'd Women

A large number of women have been involved since the first TAR festival. This wasn't by conscious design, but occurred organically. The three hosts, Pirate Emma P, Fancy Jenny Fortune and Pirate Johanna

Johanna were all women. A woman designed all of the posters and logos. Women played in the majority of the bands. All of the sound artists were women. A woman offered up the venue, and there were women technicians setting up, running the boards and striking the festival.

Radio, like many other forms of media, is still male dominated, particularly on the technical end, so it is notable that the gender balance was skewed towards women. Pirate Emma P stated, "Honestly, women are involved because a woman started and coordinates the project. There are women involved because there is a space intentionally created for women. Of course this makes a difference — it's great to create media with women, especially in a technical realm that tends to be male dominated or at least not structured so that women can participate easily."[7] Moreover, for the unruly woman there is an attraction to the life of seafaring pirates, who refuse to be regulated, and live by their own set of rules, often freely chosen by the particular ship and crewmembers involved.[8] This situation is even more transgressive for a woman pirate. She becomes a symbol for appropriating roles and lifestyles that are not hers by tradition. She plunders (metaphorically and literally) what she desires: power, wealth and excitement, breaking rules and achieving autonomy.[9] This archetype speaks to me, and the role of a woman pirate has appealed to me in the creation and running of TAR, a pirate radio station that can resurface as required, whether for protest, politics and music, or in the creation of affinity groups, festival and adventure.

I am not the only woman attracted to clandestine activities, to the freedom that comes with creative self-definition, to the liberation that comes from being a pirate radio practitioner. This capacity for self-definition is amplified because radio is not visual. It provides space for women to construct themselves without a visual focus, a space where the male gaze cannot occur. Paradoxically, radio has been said to be the most visual of mediums because of the listener's imaginative ability to construct visuals at will. As Angela Carter, known primarily for her fiction and radio dramas, contends, radio allows for magic, or the invisible, space that must be filled in by the listener.[10] This applies to all radio, but particularly to pirate radio, where there are few active constructions of women's appropriate roles, and a spirit of rebelliousness runs rampant. Listeners may still construct visuals of who might be behind the microphone, but they do so within a context where participants create their own identities, or images, while involved with TAR. These may be identities that have no real relation to who they are in their everyday lives, especially given the temporary nature of the festivals. Instead, taking part in TAR allows them to role-play, and try on different "freakuencies," so to speak.

Anonymity is encouraged and few use their real names on air. One host, Fancy Jenny Fortune stated, "We live in a culture where fully expressing one's view, especially if a woman, is not supported. Women place a lot of filters on how we express ourselves and what we express. Anonymity and the illegal nature of the event create a buffer to allow us to release opinions that might be otherwise suppressed."[11] Expanding on the appeal pirate radio might have for women, Fancy Jenny explains, "There was something about the anarchy of the whole experience. I remember there being a balance of gender — set-up, broadcasting, performance. I wonder if it links to the anarchic nature of pirate radio, and, because of the anonymity, this might be more appealing to women."[12]

As Pirate Emma P contends, it is likely that my gender and central role in TAR also encourages more women to become involved. While I look to women pirates on the high seas for inspiration and recognize the importance of role models, others may view the women involved in TAR as role models in turn. At TAR, there is a conscious attempt to welcome women in, throw them into the mix, in every capacity, technically and otherwise, in a way that is open and, hopefully, not intimidating. Historically, women have been excluded from knowledge and technical skills so that power stays in the hands of the few, specifically men. This is not the climate at TAR. Instead, it is assumed that women have competence, and can learn and develop confidence through doing, both in terms of attaining technical ability and discovering their own voices.

As women, we construct our own authority on this ship. Instead of being silenced or muted, women are asked to take up space. For many, this takes some getting used to, and it is usually a threshold to further media involvement. Pirate Emma P explained:

> [TAR gives me] a sense of being part of a media project, the empowerment of finding voice. [Events] are accessible, fun, and it's great to be part of a group push to get on the airwaves. There's nothing like taking over the airwaves and being able to be able to say what you want, however you want to say it! Ironically, the first broadcast was more of

a "help my friend out" type thing, but then led to me becoming interested in radio and going on to host and create content for community and pirate radio.[13]

TAR Continues to Grow

The second TAR pirate radio music festival took place in November 2007. We held it at the same location, and this time we had an antenna technician climb onboard. This was fabulous as she was able to fine-tune the antenna so that our range improved dramatically. This was combined with the fact that we had a friend grab his climbing gear, scale the roof and install the antenna at the peak. The difference of another 15-20 feet was significant and meant that the reception across the city of Victoria was much more solid. Often it is the placement of your antenna — rather then the wattage of your transmitter — that is of prime importance in terms of securing a solid range. Antennas work via line of sight, so the higher you can get them without buildings in the way, the better.

This time, we had 14 bands play live. (The genres included traditional, folk, alt-country, roots, jazz, spoken word, punk, hip hop/beat box and rock/experimental). Most of these bands would never get the chance to perform live on radio, and many were motivated to take part for political reasons. We used the same forms of publicity to spread the word. We once again had support from CFUV; so much so that they now requested the audio recordings from the festival to rebroadcast on the station at a later date. These audio files have been made available through websites and it means that the bands that play get radio exposure on TAR and beyond.

Despite this more mainstream recognition, it is the countercultural community that emerges with every festival that is treasured. Musicians meet across genres and develop new connections. There is a commonality of spirit in the clandestine locations where pirates hide out for the day. One of the performers, TemPest, stated, "Alternative media is rad and necessary, you hear things from local folk who otherwise don't have the opportunity to get radio play. [The experience gave me] encouragement and inspiration for sure, I loved being able to meet some local musicians I admired. Loved the random secret location bits!"[14]

Again, the eight-hour pirate radio festival was interspersed with

GARY EUGENE

*TAR pirate flag draped on parkade exterior during
2008 Victoria Anarchist Bookfair*

interviews. Each event is unique in its purpose, participants and manifestation. Coming back to the origins of TAR's name, a temporary autonomous zone is created, where affinity groups, whom are interested in direct action via pirate radio, can organize in ways that are not occurring on licensed radio stations. Groups of people who may be minorities in the mainstream media can easily take up more space at TAR. Often things are left to chance and people spontaneously show up, either through direct connections, word of mouth or internet networking. This means numerous marginalized and overlapping viewpoints can be heard over the airwaves, illustrating the diversity of perspectives within a community.

Mobile TAR

Not all of the TAR narrowcasts are as large as the music festivals described. TAR is also used to narrowcast panels and workshops, build community in neighbourhoods, and to support and publicize local events. In September 2008, TAR became mobile, narrowcasting out of a van that was easily moved and unobtrusive. We narrowcast in a public parkade, six stories above street level, in broad daylight, without interruption for five hours, in solidarity with the Victoria Anarchist Bookfair. Our Jolly Roger, our six-by-four feet, skull and cutlass banner, was thrown over the side of the parkade — it and the antenna snaking up the outside of the building. Participants at the Bookfair could not only see our Jolly Roger from the courtyard, but should the desire strike, they could wander up six floors to participate, live on-air. When we began setting up the station, security personnel were soon upon us to discover what we were doing. We were able to put them at ease with an alibi that convinced them we were doing a student film. We then proceeded to narrowcast across the city for the rest of the afternoon.

Most of the narrowcast went smoothly as radio can be relatively unobtrusive, especially when it is largely hidden inside a van. Yet live music is not as easily contained, and security personnel did circle during the performances. Three musical acts had chosen to participate in the narrowcast. The band, Gerald Fitzella, had four members present, and because it was not possible for four individuals and instruments (guitar, banjo, bass, cahon and vocals) to fit inside a van, the band set up outside the sliding door. Sound travelled, and it was during the first song that three security personnel employees were seen. Much to our surprise they simply lingered on the fringes and did not interrupt. It was clear to us that they had no idea we were narrowcasting because if they had known we were engaged in illegal activity, they would have surely thrown us out of the parking garage. Evidently they did not find the music to be threatening to the status quo, or perhaps more accurately, didn't understand *how* the music was threatening to the status quo. Upon our exit, we "dutifully" paid our parking fee for the use of the space. It was interesting to all of us that one could be fairly blatant in a pirate radio narrowcast and never be discovered. This has encouraged me to keep pushing the boundaries.

BAR TAR: Narrowcasting Broadly

Why not try to go even more public and do a pirate radio music festival out of a public location like a bar? We're not the first to create pirate radio parties. One other example was Quirk in Toronto who did all-night dance parties, inviting people to the venue or having them tune-in. Similarly he announced the events through street posters or the internet.[15] In our case, a friend pitched the idea to the manager of a working class bar that has long supported live music and local community. Surprisingly (perhaps because he didn't clearly understand what was involved) he went for it.

And so, returning to the TAR festival described at the outset of this chapter, in May 2009, a few of us arrived one afternoon around 5p.m. to set up the antenna and transmitter on the roof, ran over 200 feet of XLR cables down the side of the wall and into the club, along the ceiling of the bar to a makeshift on-air booth in a corner, and pulled a feed off of the main mixing board for the live musicians. Upon exiting the building and arriving on the roof, we were met with a stunning view of the city on a hot summer day with blue skies. Knowing that height and clearance is so valuable, our antenna technician scaled a rickety rail, in order to boost herself up onto a higher position. From here, the way was clear in every direction. We handed her up the gear, flinching a little at the sirens we could hear in the distance. We knew they were not for us, but one becomes slightly more sensitive in these piratical situations.

We had more technicians then we had ever had before, running cables and setting up, and I was left with little to do — a rarity. By seven in the evening, we were narrowcasting live across Victoria. It was the strongest range we have had yet, and it could be heard virtually across the entire city. For a city as hilly as Victoria, this was excellent for a 12-watt transmitter.

Ten bands played. Again, we had interviews, live poetry, rants and canned music between sets. A local musician and artist, J. McLaughlin, emceed from the stage as well, calling folks in to the club with mock promises of nudity on stage and rousing the listeners with comments about the importance of pirate radio in a "whitebread" city like Victoria, presumably commenting on the locality's conservative media outlets, as well as its residents. Folks played, drank, heckled, danced inside the bar and were merry. We didn't know how listeners enjoyed the suggestive humour and the tunes, but some must have, as they showed up in person with reports about the distance of the narrowcast. For myself, interested in playing with identity construction, I chose to wear a blond wig to create a sense of mystique and anonymity. Pirate Emma P also appeared in pirate garb for the evening.

By the end of the festival, past 1a.m., most folks had taken off, preferring to catch the last band or two from home. In the past, TAR festivals had occurred during the day and ended around 10 or 11p.m. Now, by the end of night, few friends were left to help out. Three of the guys from the last band to play, The Remanes (aptly named in this instance), helped to bring some gear down from the roof. My husband

and partner in crime, who is a mainstay at these events, climbed onto the high point of the roof in the dead of night to retrieve the antenna. He and I wrapped XLR cable as quickly as we could, packed up and had a final beer at the bar to toast the success of Victoria's third pirate radio music festival.

The bar festival was different in nature then the other two. Because it was in a public venue, we lost the intimacy and community building that had been so prevalent in the past. We thought that this might be replaced with a live crowd feel, and it was in part, but we had faced a conflict in our advertising. We postered the city as we had done in the past, but the venue wasn't listed because of the bar manager's concerns about liability — only the frequency. Therefore, the folks that showed up learnt about the location through word-of-mouth, primarily via the band members themselves. If we had been able to post the location freely, we surely could have had a larger crowd.

Every venue creates a different reality and brings in different people in terms of performers, technicians and other roles. Yet, there is a core group of folks who continue to support TAR, and generally appear time and again to help out. All help is welcomed. However, one of the limitations of TAR is that I am often doing the majority of the work — especially in setting up the festivals. This means that burnout is common and fewer events occur as a result. I would love to have a more committed core of people involved, or for folks to take the helm of TAR and facilitate more frequent actions. Perhaps this might happen in the future with individuals signing on to be more active in the organizing work of TAR.

Nevertheless, the temporary nature of this pirate radio station ensures that no one can become invested in a particular idea of how or what TAR should be, including myself. This is because TAR continues to change based on the venue, the people involved and the nature of the event. It also means that we can be audacious, push more boundaries and are less likely to get shut down. By the feedback received from listeners as well as participants, it is evident that TAR is appreciated and welcomed. Increasingly, more people offer up venues in private homes, in bars and cafes and in artist spaces. As options continue to expand, it is likely that the music festivals will continue, and TAR can be used for other purposes as well, such as sound art, radio dramas, protest support and, of course, community building. Temporary Autonomous Radio remains fluid, temporary and, through its autonomous nature, continues to welcome newcomers. Together, TAR pirates brashly continue to take over the airwaves of Victoria and Nanaimo. Jolly Roger and microphone in hand, TAR's creating room for new visions of radio culture and unlimited possibilities.

NOTES

This chapter is dedicated to TemPest and everyone who has helped to make TAR festivals happen, in every fabulous manifestation. It goes out particularly, to those who have offered up their time, repeatedly, so that direct action and celebration continue to flourish under the banner of TAR. Many thanks for jumping on board!

1. The kit was purchased from the Quality Kits website. http://store.qkits.com/results.cfm?CFID=7498705&CFTOKEN=21808383 (accessed May 15, 2009).
2. This kit was purchased pre-assembled from the UK, from Veronika: AAREFF Transmission Systems. Veronika offers 1W, 12W, 30W, and 100W transmitters and antennas. They have a good reputation of being reliable. Be aware that you will have to pay customs upon entry into Canada. There are a number of other online locations where one can purchase a transmitter kit or fully functioning transmitter. They vary in quality. Veronika: AAREFF Transmission Systems, http://www.veronica.co.uk (accessed May 15, 2009).
3. Hakim Bey, *The Temporary Autonomous Zone, Ontological Anarchy, Poetic Terrorism* (New York: Autonomedia, Anti-Copyright, 1991), http://www.to.or.at/hakimbey/taz/taz.htm (accessed January 22, 2009).
4. Jo Stanley, ed. *Bold in Her Breeches: Women Pirates Across the Ages* (San Francisco, California: Pandora, 1995), 20.
5. www.myspace.com/temporaryautonomousradio (accessed July 20, 2009).
6. See Chapter 8 for more on how this has been done in protest situations.
7. All of those interviewed are using pseudonyms in order to insure their anonymity. Pirate Emma P, email interview, January 14, 2009.
8. Gabriel Kuhn, "Life Under the Death's Head: Anarchism and Piracy" in *Women Pirates and the Politics of the Jolly Roger*, eds. Ulrike Klausmann, Marion Meinzerin, and Gabriel Kuhn, (Montréal: Black Rose Books, 1997), 227.
9. Jo Stanley, ed. *Bold in Her Breeches: Women Pirates Across the Ages* (San Francisco, California: Pandora, 1995), 10.
10. Angela Carter, "Preface to Come Unto These Yellow Sands" in *The Curious Room: Plays, Film Scripts and an Opera* (London: Vintage, 1997).
11. Fancy Jennie Fortune, telephone interview, February 21, 2009.
12. Ibid.
13. Pirate Emma P, email interview, January 14, 2009.
14. TemPest, email interview, April 14, 2009.
15. Carla Brown, "Pirate Radio: a Voice for the Disenfranchised," in *Peace and Environment News*, (July-August 1996), http://www.perc.ca/PEN/1996-07-08/s-

ANAIS LARUE

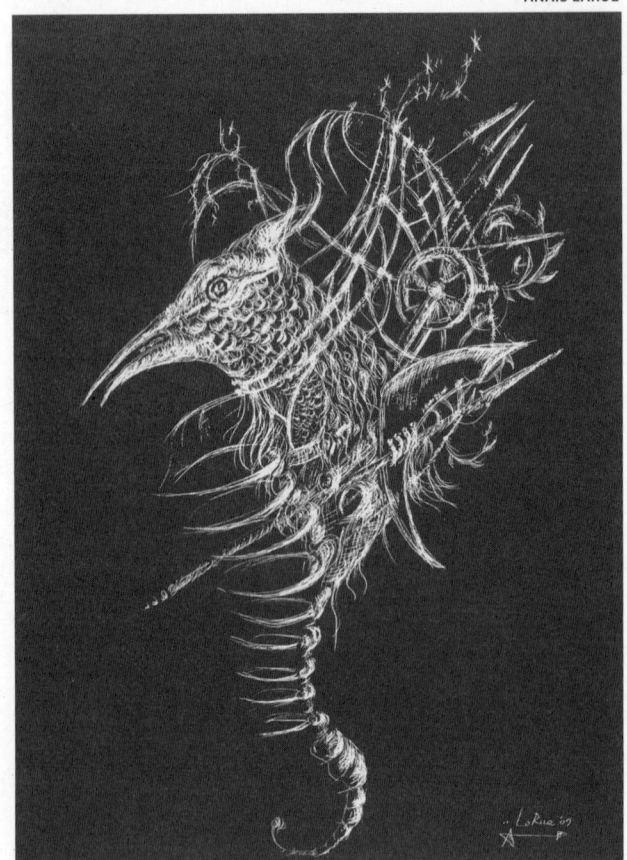

Wild and Infinite Flight

CHAPTER 10

The Voyage of a Gender Pirate and Her Toolbox

Bobbi Kozinuk

PIRATE RADIO IS ONE WAY TO CHALLENGE BINARIES IN our society. The broadcaster/listener binary was created with the growth of radio networks in the first half of the 20th century, and then was magnified even further by television. It is a binary based on the assumption of mutual dependence, whereby mass media corporations need listeners to be able to sell them to advertisers, and listeners need corporations to guide them to music and products. While it is a symbiotic relationship, the power is controlled by corporations. With pirate radio, this binary starts to melt away.

Over the last three decades, I have seen first-hand how radio can be detached from this binary and teased out across a spectrum — from corporate radio, to pirate radio, and everything in between. At the same time, I have also experienced how the gender binary of male/female can be disrupted as gender pirates question the concept of "passing" (i.e. a transgendered person "passing" as male or female), and explore how gender and attraction operate on a wide spectrum.

People are a complex mixture of traditionally labeled feminine and masculine characteristics. Within the spectrum, gender representation is fluid. Individuals are able to change on a whim between modes and show different parts of themselves to different viewers. Despite this flexibility, most people are still caught in the belief that biology determines destiny — that the existence or lack of an external sexual organ determines how one should act. They do not understand that gender is socially constructed and can be transcended. Gender is broader than biological sex, which is only based upon physical differences. Society has no room for those whose biological sex is not absolute and does not fit into the male/female binary. The imposed rigidity of this binary is more obvious in the case of intersex individuals (born with bodies that blend biological male and female aspects) in that they are usually operated on to physically define them as one sex or the other.

The gender binary is a fundamental function of a heteronormative view of gender. According to this perspective, straight men and women are seen to have a complementary mutual attraction. The stereotypical straight male needs the female to accentuate his masculinity, and to facilitate his needs so that he can do "important" things. In return, the stereotypical straight woman needs the man, to complement her nurturing role and give her life focus. Since the traditionally more task-oriented, active and tangible goals of men are granted greater value in our society than the passive and relational goals constructed for women — it remains a man's world. In examining the origins of this binary, it is clear that patriarchal society benefits from the creation and maintenance of standard gender roles. Performing normative gender roles reinforces patriarchal control. Thus, society grants power to the masculine domain and relegates the feminine to a supportive role.

Corporate society benefits from the radio binary in a similar way, by granting control to the active corporate broadcasters and usurping power from the passive listeners. To a limited extent, alternative forms of licensed public and campus/community radio can show that other possibilities exist. I liken this to the queer spectrum. Straight society begrudgingly allows queer culture to exist as long as it does not cut into its psychological and social hold on gender in mainstream society. Or, in the analogous case of licensed radio, alternative radio can exist as long as it does not cut into commercial radio's market share. Hence, the shock jocks that make fun of CBC and community radio are similar to straight conservatives who say gays can exist, only so long as they stay in their ghettos, and do not try to teach or "recruit" their kids. On the other hand, gender pirates, like radio pirates, tune into different unlicensed frequencies in order to express themselves outside of the binary and to challenge its basic assumptions.

My Journey into Pirate Radio

My journey began when I was an engineering student at the University of British Columbia (UBC) in the early 1980s, where I gained an understanding of electronics. Eventually, my interest in music brought me to CITR radio, the UBC campus station, and I realized that my electronics background would be an asset in volunteering there. After awhile, my role became more formal, as I became the chief engineer at CITR in a work/study position. During my years at CITR, I never had a radio show of my own, but helped out with various shows including a Friday night show featuring live bands. My role was a technical one — I helped set up the microphones, ran the soundboard, and did whatever was needed to make it all work. When shows were over, we, the members of our noise band Group 49, would often sit around, have a beer and say, "Wouldn't it be great if we had our own transmitter! We wouldn't have to go by all the rules! We could break free, drive up the mountain and transmit from there." I had the key to the transmitter room and we would go in there to check out the racks of equipment CITR had available to run their radio station. We would look at it all and say, "Maybe next week." It was a great fantasy, but seemed technically impossible.

Then in the late 1980s, I started volunteering at the Western Front, an artist-run centre in Vancouver that focused on video and experimental media. By the early 1990s, I was working as the technical director. It was around this time, in March 1992, that Tetsuo Kogawa, an artist, curator and media theorist who was a micro-radio pioneer, did a presentation at *Radio Rethink*, a sound art conference/festival held at the Banff Centre. Kogawa discussed his ideas about polymorphous media, micro-radio, broadcast-receiver dynamics and the laws surrounding these practices. What is more, as he spoke, his hands were not idle — they worked to build a transmitter using copper-clad board and electronic components.

Hank Bull from the Western Front saw this and was so impressed that he talked to Kogawa and convinced him to stay an extra day in Vancouver to do a presentation at their artist-run centre. I was invited to help out with Tetsuo's presentation and agreed, even though I had little information about the event. As I walked into the room, I had no idea that this day would be so influential upon the next two decades of my life. As I was there videotaping and watching Tetsuo build a transmitter within an hour, something clicked. I realized that I could do it! I could build a transmitter like Kogawa's. And in that moment, the old dream from CITR materialized.

Tetsuo Kogawa left the transmitter, plans and instructions so anyone who was interested could build one of their own. In my spare time, I started collecting parts and winding my own coils; soon, I was assembling my first transmitter. It took me three or four months to build one that worked. Soon after, other people — artists, friends, musicians and activists — heard about the transmitter and wanted to build one too. I had developed an understanding of more complicated circuits from my engineering training and once I discovered that I was able to simplify the information, I decided to start teaching workshops. After about six months, I drew a clear diagram, developed a format and assembled the supplies needed to hold my first workshop. That was 1993, and since then, I have been teaching two to five workshops per year. The workshops have been attended by a diverse group of people, including artists, activists, deejays, music aficionados and people who were just interested in technology, or more specifically, broadcast technology, and wanted the ability to build working transmitters. Also in 1993 I shaved off a big beard, started growing my hair and at Halloween went out as a woman for the first time.

Bringing Out the Toolbox: Transmitter Building Workshops

Transmitter-building workshops can take a variety of forms. They can be a presentation where I build a transmitter in front of an audience, or they can involve a work session where I engage with a small group to build a number of transmitters. Such workshops can be one day, two days or last an entire weekend. The majority of workshops I have facilitated have been in Vancouver, but I have also been invited to present across Canada, including Victoria, Gabriola Island, Hornby Island, Edmonton, Calgary, Regina, Saskatoon, Winnipeg and Thunder Bay. A particularly intense tour included a series of sold-out workshops in Winnipeg, Calgary and Saskatoon, which I called the Frozen WENR Radio Tour. WENR were the call letters I had taken on as my own, inspired by my performance art radio show with Brice Canyon, in which we had assumed the persona of the Wiener Twins, Lawrence and Sigourney, the first twins ever to be separated before birth.

A weekend radio workshop consists of two days of work, with the first day dedicated to building the actual transmitter circuit and the second to installing it into a case and building the power supply and antenna. As the participants arrive, I put them to work preparing the circuit boards — cutting up the circuit points and then gluing them down in the correct pattern. As the glue dries, I talk to participants about what we are doing and the legal elements involved when broadcasting without a licence in Canada. At this point, I assess their skill levels with soldering and solicit more information about their plans for the transmitter when it is finished. I outline possible uses for transmitters, including the strengths and weaknesses in the design of the one that we are building. One way I illustrate the possibilities of pirate radio is by showing participants a "radio station" that I built inside a video tape box. It includes a battery, a radio transmitter, an antenna that hides away and pulls out of the side of the case, and a mini-jack where I can plug in my audio recorder. With this setup I can play back sounds or speak into a microphone live. As I walk around the conference or workshop space, the transmitter sends the signal to a radio located elsewhere.

This demonstration brings alive the concept of transmission. After explaining the circuit diagram and giving a quick soldering demo, I put participants to work attaching components, starting with the cheapest and most robust parts of the transmitter and moving to the more fragile and expensive ones as they learn to solder safely. I explain the components, giving a simplified explanation of their function. By their reactions to the technical information, I can gauge how deeply I should delve. Each step of the way, I examine their solder work, having them redo bad solders before giving them the next parts. Once all the components are on, I do a final inspection making sure that they are assembled in the correct orientation. When they pass this test, I hook them up to a power meter, a sound source and a power supply. After the transmitter is working I do a quick tuning and show them that it is broadcasting across the room. Then I watch the inevitable smiles form on their tired faces. Depending on their abilities, building a transmitter can take from two to seven hours, though usually at least five. By the end of the day, participants are quite exhausted. The second day, they come back refreshed and ready to make a case to hold and protect the transmitter, as well as an antenna and power supply. One part that I always enjoy is seeing what they bring in for a case. I've seen lunch box radio, cigar box radio, and dinner plate radio, to name a few.

My philosophy with radio workshops is to make things seem simple and easy for lay people, so they can put together and build a transmitter without extensive technical knowledge. I've had people who are afraid to use a computer, program a VCR or to do anything of a technical nature eventually learn how to build a working transmitter. Their eyes light up when I test out their transmitter and it works. The excitement people seem to feel building something so technically complex is one of the main reasons why I still teach workshops after 15 years.

I get similar enjoyment from gender expression when I, with a still fairly masculine body, try to express the feminine that exists within me. My role as a mentor in regards to gender was never clearer than when I attended a program on transgender issues at the Vancouver Queer Film and Video Festival. I was with a friend and dressed quite femininely. We entered the theatre, squeezed past a young person and seated ourselves a couple of chairs away. After the screening, when the lights came up, this person turned to me and said, "You look exactly how I feel." She was obviously very troubled, trying to figure out her feelings, but not having a role model. Later when we met for coffee as friends, she told me how she would be kicked out of the house for wearing feminine clothes and had to be very discreet. She was trying to make sense of her identity and wanting to take action to advance her self-exploration. I could see that I had inspired her, and that the inspiration she was getting from me was very similar to that experienced by participants in my transmitter workshops. Though both technical accomplishments, and uninhibited gender expression may seem unattainable, both are actually possible. By modeling the authenticity of my gender expression, someone else can become aware of the complexity involved, and have the courage to live as they want.

Similarly, when I first looked at radio transmitters, they seemed impossibly complex and inaccessible — it wasn't until I saw Tetsuo creating one that I realized that such technology was within my grasp. Once this happened, I chose to share my knowledge with other people through workshops, hoping to create a comparable inspiration and excitement in others. I find this situation to be analogous to when I first began exploring my gender identity — navigating a need for a more feminine gender expression despite the limitations of a social framework where I was assigned a male gender at birth, and was subsequently raised and perceived as male. In both instances, at first I felt overwhelmed. Yet as I began to explore, and met many other trans people with different modes of expression and direction, I began allow myself the freedom to have a more expansive identity and to express my femininity. Today, by living out my true gender, I manifest a complex and non-binary expression, and hope to provide inspiration for others making parallel journeys.

Exploring Uses of the Radio Spectrum: Pirate Stations and Art

The transmitter-building workshops I facilitate have, in some cases, led to the establishment of pirate stations. The longest running is Radio Free Emily[1] at the Emily Carr University of Art and Design, which I helped set up in 1994. It stills exists as a student-run station that continues to evolve, airing music, art installations, sound design and public presentations. In the 1990s I also set up a station at the Western Front where we broadcast periodically over several years. This station even got publicity from the mainstream media (BCTV), which in turn generated interest from the general public. Unfortunately, we also got interest of the wrong sort — from Industry Canada. A few months after the BCTV piece was aired, a couple of men in suits came to the Western Front to deliver an order to cease and desist transmitting, laying out the federal regulations and consequences of broadcasting with an illegal transmitter. The station temporarily went off air, but eventually started up again to broadcast performances and other events. In addition to these two stations, I have also been invited to set up over eight others for high schools and community groups, and transmitters that I have built have been used for activist ends, such as in opposition to the Asia-Pacific Economic Cooperation Summit (APEC) in Vancouver in 1997, and against the twinning of that city's Port Mann Bridge in 2005.

Amongst all of these uses of pirate radio, the most creative has been its use in sound art. The wireless nature of transmission adds a bit of magic to performances and installations. If the receivers are hidden, the audience is not sure from where the sound is coming. A number of my students have made installations that included hidden radio receivers. From inside a sink or under a blanket, they emit a soundtrack that may be live or recorded, bringing with it an air of mystery. An example of an artist who does this is Aiyanna Maracle, a trans woman who has been influential in my own transition. While at the point of her life where she was transitioning from male to female, Maracle did a performance where she wore a radio under her clothes and told stories about her life before transitioning. Periodically, as she told the stories, a masculine voice would come from the radio, representing her inner voice. The voice would say things like "put on some pants!" or "act like a man." The sound represented the norms imposed upon her as well as her own internalized transphobia. As her friend, I knew of the struggles that she had to overcome in her transition, and when watching her perform I saw her strength. She helped me to realize that even if I could not completely free myself from those omnipresent voices that dictate absolutism in gender identity, I could at least be mindful of their power and be aware of my ability to follow a different path.

Pirate radio is a powerful tool for performances. With low-power transmission, radio can become a character in the work. The radios that Kathy Kennedy's singers carry in her large radio choral work (see Chapter 16) are an important part of the whole piece, complementing the voices in the choir. Likewise, in one of my periodic performances, *For the Birds*, I use mobile transmission to explore radio art. For the performance, I put a transmitter onto the back of my bicycle with a battery and an audio player, which supplies a soundtrack of collected bird recordings. I put a sign on the back of the bike telling people to tune into my frequency, and ride through busy bumper-to-bumper downtown traffic. The concept behind the broadcast is for motorists to tune their radios into the bird-sound station. There, they hear the sounds that would be audible if they were not encased in their noisy cars, but free to walk or cycle up the street.

Radio can also bring the audience into the work, making them an integral part of it. Low-power pirate radio can give the viewer the ability to become an antenna, a trick I have used in installations. For example, the radio receiver emits noise until the audience member positions themselves in place of the antenna, at which point the signal comes through. They can also control the reception by their movement. The receiver picks up the nearest transmitter in situations where there are multiple transmitters that are all tuned to the same frequency. In this regard, I have created several installations where people are given a radio and must move around a space to find the different soundtracks that were playing back. In *Snippets* (2006), shown at the *Interactive Futures Festival* at Open Space Gallery in Victoria, I had five different conversations playing back over five transmitters around the gallery. The listener could walk around the space and hear fragments of the different voices, creating their own mix. For *Recreational Interference* (2004), I worked with four other composers (Jean Routhier, Phil Thompson, Emma Hendrix and Michelle Frey) to create five sound works that were played back over transmitters positioned around different city parks. Listeners were given a radio receiver, a map and instructions to walk around the parks to hear the works. As they walked, they could hear the nearest transmitter until they were between two transmitters and then the signals mixed, combining two of the tracks. When they got closer to the next one, it took over so that only one was heard.

Transmitting Across a Gender Spectrum

Those who transgress the gender binary, like radio pirates, have long been relegated to outsider status. Society is deeply invested in the belief that gender is an absolute dichotomy that is indivisible from sex. Most people routinely compartmentalize, trivialize or deny their own gender transgressions. We see this kind of behavior when a woman wears her masculine-cut power suit at work, then rushes home to put on a slinky, curve-enhancing outfit to go out on a date. Likewise, the macho man puts on a flowery shirt for the beach holiday, embracing a traditionally feminine aesthetic. Yet the man with the floral shirt does not feel he is exploring a more feminine gender presentation; instead he tells himself he is on vacation and follows the appropriate script for this occasion, which includes more relaxed rules around casual attire and openness to change. When he comes back to the office, he wouldn't dream of bringing his floral shirt into the boardroom — bringing this outfit into a more traditional masculine setting would highlight its inappropriateness for a professional male. By compartmentalizing in this way, he maintains his comfort with himself as a masculine, powerful male where it matters, in the male-dominated world of business.

Just as radio can be used for much more than bringing a message from a governmental agency or private corporation to a listener, gender expression can be used for more than differentiating between male and female. We are now seeing gender labels that incorporate an awareness of non-binary gender identity: *genderqueer*, a broad term covering many non-binary identities; *genderfluid*, which includes people who may switch back forth from one gender to another, or move in other gender spaces along the spectrum; *genderfuck*, where people deliberately play with unorthodox mixes of traditional gender presentations; *andro*, where people incorporate male and female gender identities; *pangender*, which includes all genders in their identity; and, *neutrois*, or non-gendered people. They and many other trans-spectrum people see gender in new ways. These people question the supposed absolutes of our world as they move back and forth on the gender spectrum, exploring gender presentation and construct new lived realities that match their innermost being. For example, my aesthetic is generally not imposing; people need to look closely to figure me out. On first glance, people often take me for female. I often get a "Madame" until they study me a bit closer — voice, beard or Adam's apple cues are noticed, and then an uncomfortable or apologetic "Sir" follows, as if I would be insulted to lose my male privilege by being taken for a woman. Or perhaps they need to point out that they can see my "real" gender.

Dressing to display my feminine side, regardless of any perceived incongruence with my masculine physique is both an authentic gender expression and a deliberate act of transgression against the stifling confines of society. In the same fashion, my use of pirate radio attempts to question the broadcast binary by showing that anyone can use radio to express themselves in a multitude of ways. I stand in resistance to the rigidity in gender and radio, and in celebration of the beautiful spectrum of diversity. For, in the words of musician Michael Franti, "All the freaky people make the beauty of the world."

NOTES

I wish to thank my partner Jenn De Roo and editor Andrea Langlois for their invaluable help in writing and molding my notes into this essay.

1. More information about the station's current programming is available on the Student Union's website at: www.emilycarrstudentunion.ca/index.php?section_id=15 (accessed August 8, 2009). When I was setting up Radio Free Emily, I approached the CRTC and asked what would happen if a student were to use a home-built transmitter. The representative to whom I spoke hemmed and hawed and finally said that if there was an interference problem they

would let us know. In my reading of the Broadcast Act, there are exemptions for low-power transmitters, as long as they are stamped as certified by Industry Canada, which is a long and costly process. In reading their literature, I found an exemption for "Certification for Home Built Transmitters, Not From a Kit," which I consider to be an exemption for the kind of transmitters I build.

PHOTOMONTAGE: PHILIPPE / STILL PHOTOGRAPH: JEFF

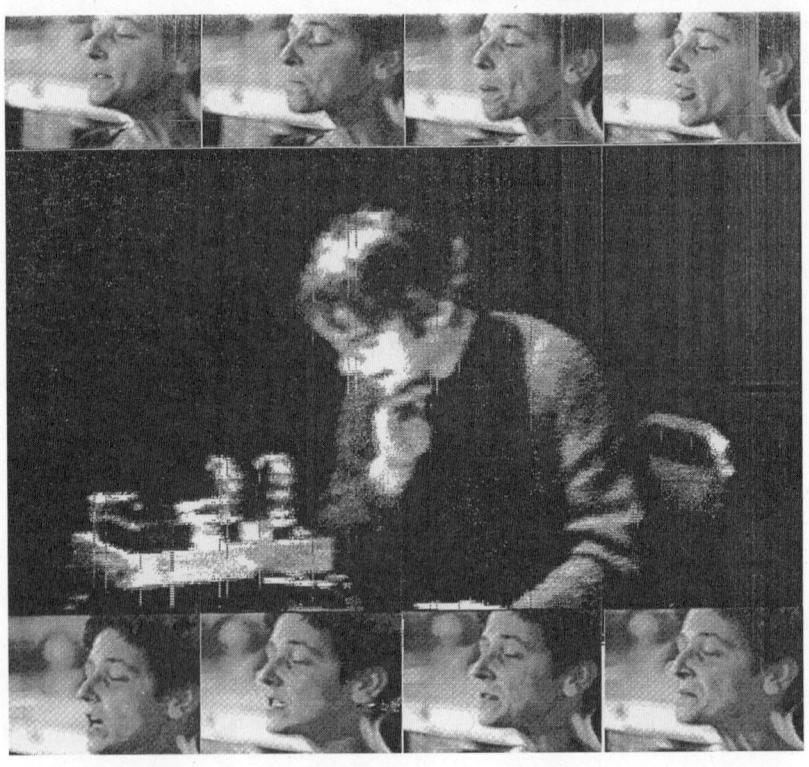

CHAPTER 11

Pirate Radio and Maneuver
Radical Artistic Practices in Québec

André Éric Létourneau

WHEN ART IS PRACTICED WITHIN THE CONTEXT OF unauthorized broadcasting, radio space becomes the place and medium of a radicalized art. This chapter will examine how pirate radio broadcasting falls within the parameters of a specific art practice — the "maneuver." Here, the term maneuver refers to an artistic act that is interdisciplinary, immaterial and, for the most part, has a limited visible component. Stemming from the performance art and happenings of the 1960s and 1970s, the maneuver seeks to dematerialize the art object to the point of near total interference with time, space and the environment by placing the action, rather than the artist, at the centre of the process. A maneuver occurs without permission within society, geographical space and lived experience, and is almost entirely intangible. This disembodied practice — broadcasting within public space without obtaining permission — makes it a perfect partner to pirate radio. As a result, the maneuver often appears to be a social experiment in which participants interact within the context of an art initiative with no tangible artwork, no passive spectator and no individual artist. In effect, the public becomes co-creator of the artwork.[1]

Even though the maneuver has various antecedents in the sphere of art, its theoretical foundations lie in the work of the artist Alain-Martin Richard, a member of Québec City's collective *Inter/Le Lieu* in the late 1980s.[2] Richard continues to work in large-scale maneuver projects that resonate within social spheres inside urban and rural settings and online. Within this essay, the concept of the maneuver is applied to interdisciplinary initiatives in which illegal radio broadcasting becomes a prevailing factor in the process of interfering with social reality. Pirate broadcasts produced within the context of a maneuver follow forms of art practices found in the tradition of experimental art production[3] and are rooted within radical social movements.[4] The maneuvers discussed herein were undertaken in Québec between 1983 and 2005 by different artists and collectives.[5] Pirate radio broadcasts produced by these artists were for the most part "substitution broadcasts," meaning they temporarily jammed the broadcasts normally issued by commercial radio stations within a limited geographical area. Their attacks against licensed radio stations had a guerrilla nature and aimed to temporarily occupy their assigned space so that the audience member listening to the programming on that frequency would have a unique experience. The execution of these works thus took into account the regular programming broadcast on the commercially "owned" frequency. A substitute broadcast replaces the music usually heard with new aural/oral experiences. The content of these substitute broadcasts often embodied a critical position against the regular programming heard on the pirated frequency. Other broadcasts explored here took place within radio's free spaces — the spaces between used frequencies, which are unassigned and therefore available for pirating. These broadcasts are what I call "supplemental broadcasts," which tend to be community-based and festive.[6]

Parasitizing

Since the early days of radio, its producers have tried to create broadcasts within a restricted, highly fragile and easily disrupted space — the airwaves. Every station uses a single sensitive frequency to transmit inside a given geographic area (except in some totalitarian States where the airwaves may, in some cases, be occupied by a sole broadcast — or sole noise — intended to jam foreign broadcasts).[7] Like still waters, compromised space can be disrupted, and the shores of frequencies are susceptible to opportune interferences by those who describe themselves as "parasites." Claiming the socially denigrating term of parasite as their own, radio pirates have continued to muddy the waters. By using a process of "parasitizing" the airwaves, which is akin to the self-regulating processes of the natural world, deviant broadcasters reestablish an organic balance between institutions and marginalized groups. From its very beginnings, radio was developed to deliver clean, tidy and clear messages, in order to open a space for the State and industry. It was yet another territory to colonize. As Saint-Thomas l'Imposteur remarked in an interview,

> Stations multiply and provide aural landscapes dedicated to a virtual sense of being at home, made portable via small listening devices like Walkmans and transistors. The broadcast starts from point A and moves towards, through geographical space, multiple point Bs. The thinkers behind radio production intended the space between the two points to be clean, free of intruders and parasites. War is waged against interference.[8]

Pirate Radio as Paradigm and Social Construction

It may be because of the romanticism with which pirate radio is imbued that the term pirate is commonly accepted by those who engage in the activity. Yet it is important to note that the term also implies criminality, and it is worth questioning its usage. When attached to a radio broadcast, the term pirate carries a specific dramatization of an act involving the reappropriation of the original freedoms that existed before radio became monopolized by government and corporate institutions. Of course, the term is inspired by the off-shore situation of those stations broadcasting illegally from ships in international waters. If pirate radio sometimes finds refuge on ships, the analogy "empire of the airwaves"[9] is certainly relevant. To associate an act of illegal radio broadcasting with the act of piracy also underscores the illegal aspect of these activities, reminding the violators, on the one hand, of their social marginality and, on the other hand, affirming, at least among mainstream media's audiences, the illegal aspect of the act vis-à-vis the privatization and nationalization of radio frequencies. In other words, the term pirate, when attached to radio, can act as both a tool of dissuasion and as power of persuasion.

Radio as social paradigm is intricately linked to the notion of space, especially with regards to its privatization and its access by institutional powers. Because the number of frequencies on the airwaves is technically restricted, State radio broadcasting — then private radio

— delimited this space by *signifying* it. Borrowing an idea developed by Gilles Deleuze and Félix Guattari,[10] in relation to the training of educators and applying it to the *signifying* of spaces, a pirate broadcaster does not pillage the radio space as a "real" pirate might, but simply *uses* it. In this usage, the pirate broadcast interferes with the symbolic *signifying* of radio space — with the meaning of what it is and what it can be. Such radio piracy de-signifies radio as an institutional space, and re-signifies it as an *other* space, a citizens' space, not one of the noisy majority, but rather that of marginal populations whose voices are silenced by the roles imposed by the dominant class and by institutions. The foundations of commercial radio rely on the signification of space as an economic territory targeted for profit. Pirate radio, as we shall see, challenges these very notions.

Angry Thérèse: Owner of the Building Housing a Pirate Radio Station

It is only natural then that the content broadcast through pirate radio is often a commentary on the tensions between the dominant social classes and marginalized groups. The first example used to illustrate this is the broadcast "Thérèse," on Pouf-FM. Active between 1983 and 1990, the pirate station Pouf-FM broadcast from Québec City's Haute-Ville quarter of Saint-Jean Baptiste. Promoted among its network of fans with the ambiguous slogan "Pouf-FM : le son au maximum" [Pouf-FM: Maximum Sound], the station was operated anonymously by one man, Pouf, who acted as deejay and broadcast a variety of alternative music rarely ever heard on commercial airwaves.[11]

But the noteworthy importance of this station's programming went beyond music. Once a month between 1983 and 1987 (and then sporadically until 1990), Pouf-FM broadcast a recording of approximately 10 minutes of a conversation between the owner of an apartment and her tenant. The two argued over a legal dispute, each contesting the others' increasingly virulent retorts, as the discussion regarding a late rent payment quickly escalated into a quarrel between the two protagonists. Each time, the recording ends at a climactic point when the voice of the furious woman shouted repeatedly, "I want my money! I want my money!" before slamming the door as she left the room. The recording was not staged. It represented an indisputable fragment of the real.

POUF'S BROTHER

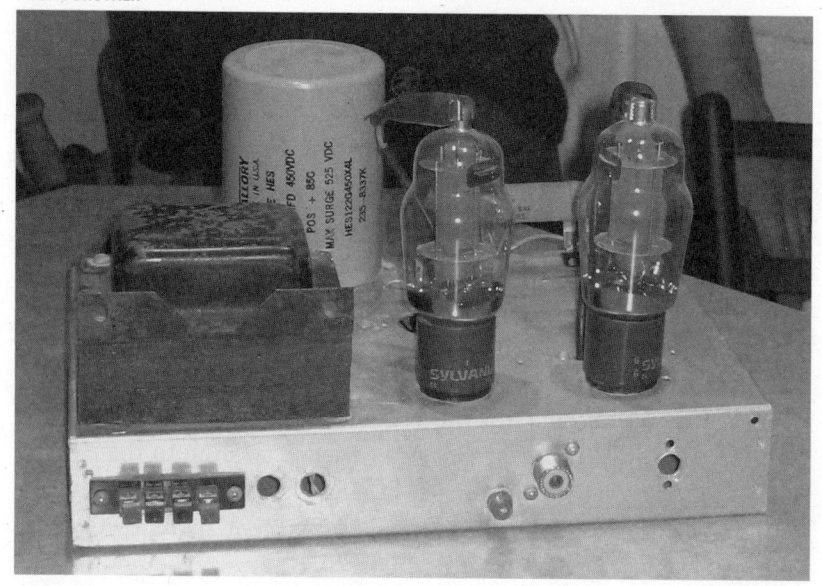

FM transmitter built by Pouf

"Thérèse" was aired any time a listener called in a request to hear it. The popularity of the piece was due to several factors. The situation irreverently represented the conflict between social classes (landlords and tenants) and/or between the pre-war and post-war generations. Aired in a pirate radio context, the recording also became a metaphor of resistance against privatization — the appropriation of the airwaves by a privileged segment of the population with economic power and influence over the management of public areas. This is perhaps why listeners regularly phoned in to request the amusing recording. The effectiveness of the piece was thus increased tenfold by the context of pirate broadcasting, which pushed irreverence to the point of near indecency. This initiative was directly linked to the community and was often aired at parties thrown by Pouf-FM listeners.[12]

Although this initiative didn't consciously subscribe to an art process at first, it constituted a compelling introduction to the radio maneuver. Moreover, the transmitter built and used by Pouf became, in 1992 and 1993, the centre of maneuvers carried out by the collective algojo) (algojo in the context of Interzone, an official art initiative event presented by Le Lieu, an artist-run centre in Québec City.

algojo) (algojo, Jeff and Voxain

Established in 1990, algojo) (algojo originally consisted of a team of two artists interested in inserting art into the context of radio broadcasting. It became one person's project between 1993 and 1997, and algojo) (algojo created performances and *hörspeil* (radio plays) that linked radio space with the practice of public performance. The pirate element of the duo was conducted by Jeff who, along with Pouf, modified Pouf-FM's original transmitter to enable the splicing of the electric current supplying the apparatus to the body of the performer who was directly connected to the power source.

The frequency of pirate broadcasts varied according to the electric shocks Jeff inflicted upon himself during the public performance.[13] Jeff considered this device a musical instrument in its own right. The source of the sound of this instrument lay in the connection between the sounds produced by the performer and his body, and the sounds already existing on official airwaves. He called this device "Voxain." Voxain was in fact an FM transmitter illegally broadcasting the performer's voice on the radio band while cyclically changing the frequency. In effect, the performer would give himself electric shocks with the transmitter's lights. His body was directly connected to the transmitter's electric current, and the two entities became one. Because the intensity of the current had an impact on the broadcasting frequency on the airwaves, current variations caused by the sharing of electric current between the performer's body and the radio transmitter caused the broadcast to travel throughout the airwaves. The greater the current, the more the pirate broadcast jammed the lower frequencies of the airwaves and the more altered the timbre of the performer's voice became as his nervous system absorbed the electric shock while he continued to read from a script. In the performance room, the result was heard through about twenty small radio receivers, each of these tuned to a different frequency. Thus the voice of the performer, which travelled throughout the airwaves via electric charges, could be heard at the same time that official radio broadcasts were momentarily jammed.

For a listener tuned into a specific radio station frequency, the performer's "parasitic" voice was heard through Voxain by sporadically crossing — for a few seconds — all frequencies transmitting regular broadcasts. Listeners tuned into a commercial station would therefore hear, for a few seconds, fragments of texts that were inserted, more or less randomly, into official broadcasts. The 1992 *Festival Interzone* organized by Le Lieu, *Homélie amplitudes A1 A2 A3 A4 fonction du temps* (Homily amplitudes A1 A2 A3 A4 function of time), showcased the algojo) (algojo duo in a performance that lasted several hours in which the subjected body's exhausted voice passed through the transmitter. The performers thus appropriated all of Québec City's Basse-Ville stations to make them the space where random texts were interspersed within the programs of all regular FM stations. According to Jeff, "The listeners at home certainly encountered incomprehensible interruptions, at the same time that they experienced — through the random aspect of each person's experience — an ungovernable artistic experience with qualities beyond our control."[14]

The artist's words above summon a theory I proposed in 1991 — that of "polysynesthesia." The term polysynesthesia is a neologism I created[15] to describe the phenomenon experienced by the radio art listener. It describes the overlap of different sensorial impressions that the listener experiences during the radio transmission, the overlap of permeating sensations, often random or in synch with other events. Each listener experiences a different phenomenon upon listening to a radio work. The work, or at least its reception by the listener, thus escapes the artist's control and moves into multiple settings with random parameters. This idea is meant to describe different interconnected sensorial experiences during the listening experience. Multiple uncontrollable factors interfered in the execution of the work and, in a way, played the role of co-author in the performance of *Homélie amplitudes A1 A2 A3 A4 fonction du temps*. This maneuver was based on the idea of interference with existing programming, parasitizing or free-riding or temporarily occupying the FM band without a licence. It was a genuine critique of the fetishistic system surrounding pop music, which is subject to the fickleness of the economy and fashion,[16] that which the Québec composer Pierre Mercure incidentally described as *"mauvais goût ambiant"*[17] (ambient bad taste).

Transmission Lesson: Aural Erasure and Hijacking Radio Broadcasts

The act of using radio as a "parasitizing" process against the institution of official radio was also played out in another maneuver carried out in Montréal in 1992. The *Leçon de transmission* ("Transmission Lesson") of Karma Terraflop again used Pouf's transmitter, and was carried out on Mont-Royal Avenue in Montréal. *Leçon de transmission* was aimed specifically at listeners in cars travelling inside the area covered by the pirate transmitter. Terraflop described the maneuver in an interview conducted in 1995:

> At the time of newscasts on the major radio network of Cité Rock-détente, I substituted the broadcast with a choir made up of about forty voices randomly yelling and screaming "Help!" for several minutes. This went on for awhile as the listener of Cité Rock-détente drove through the pirate broadcast area and heard the overlapping stations, followed by the substitution of the original transmission with my own broadcast, which is another aural space carrying critical overtones... It was almost as though the two stations conversed in the overlap. It was my statement on the quality of the regular programming of major radio stations: a constant noise broadcasting music and information with content that bowed down to the establishment and intended to distract peoples' attention from relevant social issues. In this way, the commercial music industry is like an enemy, because it often maintains this alienation by imposing syrupy and commercial music on radio broadcasters to the detriment of relevant news stories or documentaries, or even non-commercially oriented music. The same goes for the newsrooms of the major media outlets that camouflage important events in order to broadcast "masked" information that rarely concerns the everyday lives of people. The truth of collective progress is hidden behind the incessant "noise" of commercial radio and major media outlets. And community radio, even though it tries to compensate for this situation, is itself structured by social conditioning, and watchfully regulated by corporate culture. For example, the editorial content of community stations in Montréal is often drawn from the major newspapers, namely *La Presse*, which is part of Power Corporation. Pirate radio is a way of distorting the established order. By substituting regular transmissions, I temporarily activate aural *erasures* of the semantic garbage all around us. Perhaps it's a type of revenge aesthetics. It's a tiny dose of parasitizing, a defeat before Goliath, but I nevertheless recast the space for a time, without getting caught by governmental control systems, the police of the airwaves.[18]

This quote eloquently elucidates Terraflops's apprehension about the response of State mechanisms for the control of the airwaves. For example, while Pouf-FM was in operation, it eluded (on several occasions) detection by the electronic devices of the State, which, when alerted by unknown parties, are used by CRTC (the Canadian Radio-television Telecommunications Commission) personnel to uncover transmission sources. This fear of being located has led most pirate broadcasters to develop specific strategies related to the length of the transmission or to the tracking used by government vehicles dedicated to locating pirate antennae.[19]

Radio Art in Institutional Spaces with FM Micro Broadcasting

At about the same time, I carried out, along with Québec City's *Diffusion système minuit* (Midnight System Communications), various maneuvers alongside several artists discussed here, who used radio as their medium in the context of a radio art and performance event. Over six days, the interdisciplinary event *Retour de voyage en ces lieux oubliés de l'anéantissement* (Back from a Trip to Annihilation's Forgotten Places) presented, alongside an archival display of radio creations, over fifteen performances, namely those of Jeff, Karma Terraflop and algojo) (algojo, but also those of interdisciplinary artists Christof Migone, Willem de Ridder, Luc Desjardins, Neil Wiernik, Emmanuel Madan, in addition to the aboriginal sound artist Chris Wind and Québecois electroacoustic composer Michel Smith.[20] The audience was equipped with various receiving devices, which enabled them to listen to the exhibition on the FM band throughout the exhibition's space and the perimeter of the building housing Articule, a Montréal artists' centre, where the event unfolded.[21]

Sonia Pelletier, art critic for *Inter* magazine, described the event as follows:

> Back from a Trip to Annihilation's Forgotten Places provides an unconfined visual framework tuned into radio works (systems of hidden tape recorders, earphones, documentation transferred to videotape), pirate transmitters (two frequencies), and olfactory, tactile and visual evocations. Radio pieces were selected from Indonesia, Holland and Canada. The feat of such an initiative lies, in my opinion, in the execution of the idea that art coincides with the assertion that the essential is found in the invisible, and that process eludes time. In this

installation, the "communicable" resides in the travelling, the vehicle, the trajectory, the movement and the transformation.[22]

The four transmitters used in this event allowed the radio works to be listened to not only in the gallery, but also throughout the intersection at Mont-Royal Avenue and Saint-Laurent Boulevard. The entire intersection was supplied with four new stations broadcasting pre-recorded works for over a month, as well as contextual works with elements directly linked to the neighbouring space (Le Bouchon restaurant in front of the gallery, Mont-Royal Avenue and the involvement of passers-by on the street). The event fit naturally into the logic of an art exhibition, only the gallery's space extended past its walls. The gallery, located at an undisclosed location at the intersection, thus displayed itself audibly in the neighbouring urban landscape.

Abribec and the Mobile Duty-Free Zone

In April and May 2003, Abribec, a collective with a socio-political bent, used a weak radio transmitter in its maneuvers, which were carried out within popular demonstrations organized by the *Collectif Québec sans-pauvreté* (The Québec Anti-Poverty Collective). Abribec, a group of artists executing maneuvers around the issue of tax shelters for institutions and the economically privileged, originally emerged from an initiative involving a series of posters in various Québec City bus shelters, which are overseen by the advertising company Viacom. Thanks to the financial backing of art organization *Engrenage Noir*, the posters denouncing big-business tax shelters were displayed for an entire week. The poster's design parodied the Québec government's Agir (Act) campaign, which promoted public works subsidized by the provincial government of the time. The poster presented the following text:

> ACT against the inequalities and prejudices directed at people living in poverty! Against employers who seek tax shelters for themselves but refuse to raise the minimum wage above poverty levels; against soaring profits but no income security threshold to cover basic needs. What kind of future do we want? A responsible society would say, "Enough with double standards!"

At the bottom of the poster was a Québec flag — symbol of the government and the Québécois nation — violently disfigured by the dan-

ABRIBEC

CAMPAGNE DE SALISSAGE FISCAL

www.abribec.qc.ca

CAGIBI INTERNATIONAL   ISO 1000000

AbriBec©
Suppôt de la nouvelle humanité fiscale

Abribec stickers handed out during the maneuver

ger-of-explosion symbol found on household aerosol products.

The week after the maneuver, the Cossette agency, in charge of the government's promotional campaigns, asked Viacom to remove the Abribec posters from all bus shelters in the greater Québec City area. They stated, "The request comes directly from the Premier's Office, which deems the poster misrepresentative and blames its authors of distorting the message."[23] Commenting on the Abribec maneuver in the weekly paper *Voir*, journalist Tommy Chouinard retorted, "In 1999, 103 firms made $11.3 billion in profits, but only paid taxes (federal and provincial) amounting to $394.5 million, which represents a real tax rate of 3.5%! Some 33 businesses recorded earnings of $1.8 billion and paid not a single cent from their coffers in taxes, while 158 companies posted net profits of $25.6 billion dollars and paid only 10% tax."[24]

The sequel to the Abribec maneuver was performed within a demonstration organized by the militant *Collectif pour un Québec sans pauvreté*. A mobile "duty-free zone," delineated by the area of a pirate radio transmission, travelled inside a sculpture on wheels that represented a mobile tax shelter. It took up about two square meters and the transmitter covered an even larger surface area, creating spaces within which it was declared that citizens were exempt from paying taxes to the State — just as large corporations were through tax shelters. Over an area of several dozen metres around the moving sculpture, the Abribec pirate transmitter broadcast the following message, interspersed with silence, delivered by a colourless voice reminiscent of the type of voice heard in airport duty-frees and Radio-Canada announcements: "You are in a duty-free zone. Cagibi International would like to remind you ... you are now in a duty-free zone."

The duty-free zone moved around in this way at a demonstration by community organizations and political groups headed towards the

National Assembly (Québec's parliament), and made its presence felt on Abribec's transmission on the FM band. The broadcast entered the neighbouring houses, then around and right in front of the National Assembly where demonstrators congregated. Stickers reading "*campagne de salissage fiscal*" (fiscal dirt campaign) denouncing corporate tax abuses were handed out by members of the collective. The maneuver was based on the combination of Abribec's shock imagery (a culture jam based on the government's graphics-based campaigns), pirate radio and direct and concrete actions. After the demonstration, Abribec's radio transmitter was handed over to the *Squat de la Chevrotière* (Squat of the Chevrotière), located right in front of Complexe G in Québec City's Haute-Ville.[25] The squat then used the transmitter in its operations involving protection and an information campaign to curb condominium construction in the historic neighbourhood. The squat was dismissively shut down months later by municipal authorities following a city-issued eviction order.[26]

Conclusion

The aforementioned examples may give the impression that government enforcement of regulations concerning the airwaves in Canada is rather lax. Indeed, most of the examples referred to in this book have broken — even if but slightly and occasionally — Canadian radio broadcasting laws. If these examples have been written about in a way to keep their players anonymous, it is clear that this is due to the fact that describing their actions raises the legal, ethical and socio-political stakes. And although some of these maneuvers have been written about in other publications, and others were discovered and brought before the authorities, there have been virtually no legal decisions rendered against them in cases charging them with breaking the laws governing the empire of the airwaves. Does the Canadian government show leniency towards such minorities who stealthily or slightly infringe its radio broadcasting laws? Or, does the Canadian State prefer to carry out its enforcement activities undercover, a strategy to minimize any disclosure of the actions taken by groups that challenge the powers that be? Wouldn't creating hoopla around a radio scandal run the risk of promoting some ideas put forward by marginalized groups working for a more "Just Society"[27] and critical of the economic domination of public powers and culture? Or, does the act of transmitting in the name of art represent a lesser wrongdoing than an unlicensed radio transmission with strictly political intentions because of the "impunity of art,"[28] which makes the wrongdoing a cathartic and liberating act of maneuver? Or, do radio pirates simply make a limited impact that does not worry State-run institutions controlling public spaces, leading these institutions to tolerate such intruders rather than vigourously prosecute them?

The answers presumably lie in the delicate balance between the possible responses to each of these questions. Pirate radio falls within a landscape peppered with a plethora of commercial broadcasts that dominate the airwaves with their strong signals and large areas of transmission. Pirate radio only somewhat disturbs the balance already attained by radio stations. In this proliferation of symbols, of *signifying*, pirate radio leaves tiny traces. The sterilization of the airwaves is not — for now, in any case — put at risk by this form of resistance because of the concentration of media in the hands of a limited number of conglomerates.[29] This kind of media monopoly is characterized by problematic situations, such as the restrictions placed on the types of news stories chosen by those at the helm of public radio stations,[30] the increasingly hard-to-find free spaces on radio stations filled with formatted-radio content, and the repetition of news stories of major media outlets on community radio stations. In such a context, radio pirates and artists operating with unlicensed transmissions, whatever their format continue to raise the stakes in an exemplary manner by providing models of radio use that are radically different from those promulgated by dominant institutions. The relationship between established, licensed radio stations and pirate radio broadcasts creates a fertile breeding ground for challenging the economic reign over public spaces and the airwaves.

This phenomenon also questions the impact of neo-liberalism and the players in radio whose values are reflected in their standard programming throughout the mediascape. By extension, this dynamic reminds us that practices commonly accepted in society can be questioned and challenged, and that perspectives have to be multiplied in order for socio-cultural contexts to evolve. In this sense, the radio maneuvers discussed in this chapter serve to highlight these issues and effectively expose the unnecessarily functional format that makes up today's radioscape.

NOTES

Ths text was translated from the French by Clara Gabriel.

1. Some ideas here are borrowed from Stephen Wright, and were developed in his "Vers un art sans oeuvre, sans auteur, sans spectateurs," in the catalogue of the XV Biennale of Paris, Éditions Biennale de Paris (2007).
2. Alain Martin Richard, "énoncés généraux-Matériau: manoeuvre," *Inter*, #47, Québec: Éditions intervention (1990), is, in my opinion, the seminal work that defines the exact characteristics of the maneuver.
3. Howard S. Becker, *Art Worlds* (Berkeley: University of California Press, 2008).
4. A maneuver often becomes part of an environment and the everyday acts of being. Although the term qualifies practices that for the most part take place within urban space, the term here is applied to a certain type of intervention that uses unlicensed radio broadcasting — in other words, pirate radio — as a central element of the art initiative intended by the maneuver.
5. Many of the pirate radio practitioners and artists discussed here use pseudonyms or artist names in order to protect their anonymity.
6. The methodology applied in this essay highlights the junction between social science publications and a series of interviews conducted between 1992 and 2009 with various operators, radio broadcasters, and artists engaged in maneuvers and unlicensed broadcasting.
7. A piece by Italian composer Walter Marchetti, a bruitist mass of sound made up entirely of noise created by the Francoist censorship in order to jam foreign airwaves throughout the Spanish territory, would have deafened John Cage for hours following a concert performance. Marchetti's approach represented a unique response to the State's attempt to control the airwaves.
8. Saint-Thomas l'Imposteur (pseudonym used to insure anonymity), personal interview.
9. French literary term designating ocean, used in my essay "L'empire des ondes," in *Parallélogramme*, vol.16, no. 4, Éditions ANPAC\RACA, Toronto (1991). The term was reused in 2007 by journalists Aymeric Mantoux and Benoist Simmat in the title of their study on the French network Radio NRJ.
10. Gilles Deleuze and Felix Guattari, *Mille plateaux : capitalisme et schizophrénie* (Paris: Minuit, 1972).
11. Pouf's transmitter, FM and mono, was built along the lines of the classic tube transmitter and produced a 15-watt transmission.
12. Pouf (pseudonym used to insure anonymity), personal interview, June 1992.
13. A drop in electric voltage causes the broadcast frequency to go up to 108 MHz on the radio dial, and a return to voltage causes the broadcast frequency to drop to the original frequency (about 89.5 MHz).

14. Jeff, artist, interview, 1994.
15. André Éric Létourneau, "L'empire des ondes," "A World of Waves," in *Parallélogramme*, vol. 16, no. 4, Éditions ANPAC/RACA, Toronto (1991).
16. Theodor Adorno, *Le caractère fétiche dans la musique* (Paris: Éditions Allia, 2007).
17. Pierre Mercure, in *Musique du Kébèk*, ed. Raoul Duguay (Montréal: Éditions du jour, 1971).
18. Karma Terraflop, radio artist, interview, September 11, 2001.
19. Pouf, personal interview,1992.
20. Smith's work bore the title "Le cimetière des ondes radios."
21. Marie-Michèle Cron, "Cris et chuchotements," *Le Devoir*, Montréal, October 1992.
22. Sonia Pelletier, "Ondes fluides et points de force," *Inter*, no. 55, Québec City: Éditions Interventions (1993).
23. From Abribec's online archive, www.iso1000000000.ch/abribec/ (accessed May 23, 2009).
24. Tommy Chouinard, "Tous aux abris... fiscaux!" *Voir*, Québec City, July 11, 2002.
25. Complexe G, a 31-story government building built in 1968, houses numerous important Québec government offices. It is the tallest building in Québec City and dominates the skyscape.
26. The adventure of this squat, whose purpose rose essentially from the fight for social housing, is to this day commemorated yearly by various citizen groups of Québec City, Hommage au 920 de la Chevrotière, Centre des médias alternatifs du Québec, www.neonyme.net/squat/cmaq/index.html (accessed August 6, 2009).
27. "Just Society," was the famous electoral slogan of Pierre Elliott Trudeau.
28. According to Jacques Soulillou in his *L'impunité de l'art*, acts that transgressed or were even labelled deviant would be more socially acceptable, if not encouraged, in the contemporary art world rather than in most other sectors of human activity. Jacques Soulillou, *L'impunité de l'art* (Paris: Seuil, 1995).
29. Ben H. Bagdikian, *The New Media Monopoly* (Boston: Beacon Press, 2004).
30. This problematic is brilliantly described by Mario Gauthier in his "Standard III ou Un silence en cache-t-il un autre?" in *Standard III*, a booklet accompanying Benjamin Muon's double-CD, Éditions PPT \ Strambogen, Paris, 2009.

CHRIS McCLAREN

Absolute Final Intimacy of the Ear

CHAPTER 12

Touch That Dial
Creating Radio Transcending the Regulatory Body (1990)

Christof Migone

The Project of Communication is Borne of
a Passion for Creation

WE HAVE ORGANIZED EVERY SECOND OF THE AIRWAVES into categories. Everything is something and nothing is left uncertain. And if Bertolt Brecht could say that radio is one of those inventions nobody ordered, its realization now always seems to occur in perfect order. The verbal adroitness — the deadly fluency of the trained voice — formats our listening and provides us with standards which shape our ears into solitary and passive frames. By creating undefinable waves, radio art de-tunes our conditioned frequencies. Disoriented, we find that allotted frequencies have restraining contours and we opt, rather, for the static of non-broadcasting frequencies. In that ether we find the veritable potentiality of communication without its power signifiers. Yet, it is in the physicality of existing radio studios that radio art is created and aired. This implies an ever present conflict between the medium in its endless phase of justification and the artist whose subjectivity may confront, resist and pervert confining regulations (imagine a musician forbidden to play certain notes). Radio artists in Canada have the distinct pleasure of having the Broadcast Act and Canadian Radio-television Telecommunications Commission regulations as their palette (colours included and always balanced). Other states, other regulatory bodies, same pre-fabricated palettes. The airwaves, akin to airspace, have always existed under the imperatives of national interest. From Hitlerian fanaticism to the biased objectivity of the Voice of America, the airwaves have often been used more to proselytize than to communicate.

The Project of Participation is Borne of a Passion for Playing

Tuning (in) is an act of defining amongst a slew of pre-determined approaches to radio. To touch that dial is a seductive gesture that implies not only the plurification of the available choices but the mapping of choices hereto uncharted. The outcome will invariably challenge today's singularly mundane credo based on one speaker at one microphone: a single voice addressing the audience directly, an invisible power, a one-way channel, an authority excited by its technological medians. Unlike breath, radio's one-way power message dissipates energy. To create radio we need to accomplish the impossible, we need to de-mediatize the medium. We need to strip radio of its seemingly inherent need to legitimize itself. In the electromagnetic spectrum, FM broadcasts overlap with television transmissions; an analogy for audio's subservience to visual representations (where imagination is imagined for you). Radio, after all, is dark; it is tied only to fragments and then scatters. The art of radio does not necessarily assume coherence, it assumes that composing in disparate juxtapositions can create new, manifold relations. The decisive step is left to the listener. In the mind of the listener the fragments of meaning will come alive. That is the first step toward making the listening participatory. The listener, however, is not asked to become a barometer of approval, but rather an artist in turn. Without the dissolution of the distinction between both roles, we still appear separate, divided, either here or there. And with radio, I would rather create within the intimacy of an ear than be stuck apart with geographical distance. The text of radio art is signed in sound. The text is authored by those who breathe it. An attempt to bypass dualities and activate a collective authoring of the air. Radio art is the technology of breath. We are all receiving all stations at all times. Like an involuntary muscle, we are breathing each other.

The Project of Self-Realization is Borne of a Passion for Love

A skipping record, the wrong turntable speed, dead air: deejays' worst nightmares are the most common compositional tools for the radio artist. Stutters, burps, hems and haws. Excess (through these "mistakes") is a necessary stage for live radio artists in purging themselves of the myth of the radiogenic. The world of radio is populated by fences, imaginary fences of quality. This demarcated territory can be reappropriated by an act of playful measuring, categorization mutates to become fractal geometry:

> bellybutton to nose: 24in.; neck extended: 7in.; mouth to ear of other: instant lengths; transmitter to transmitting: 47 abrasions, 32 bruises and a light concussion; right knee to hip: 20in.; birthmark to tattoo: 5in.; breathing each other: 2 or more radio plagiarists; chest to chin: 11in.; depth of listening: no. of captive kilometres.

These contorted measures are determined by the imagination and exemplify the evasive parameters of radio art. Through them we get a glimpse at radio's tenuous and shifting contours. We are mapping the body of a medium whose form is based on a language of otherness, displacement and transmission.

Isolated, the Three Passions are Perverted

One, the project of communication is borne of a passion for creation. Two, the project of participation is borne of a passion for playing. Three, the project of self-realization is borne of a passion for love. Isolated, the three passions are perverted. Dissociated, the three projects are falsified. The will to communicate becomes artificial objectivity; the will for participation serves to organize the lonely in a lonely crowd; the will for self-realization turns into the will for power. The homogeneous radio landscapes sanctioned by formulae comprised of lowest common denominators have unfortunately deadened the voices of the airwaves. Yet despite the medium's persistent dead set monotony, in sporadic pockets of creative resistance, an art of radio is being conceived through a process of deconstruction, demystification and deformatting. Once stripped naked, rather than dictate sense, radio

Radio Naked

Tactics Towards Radio without Programming

By Christof Migone

1. Always give the wrong time, date, weather and news report.
2. Constantly change your broadcasting frequency.
3. Do any technical repairs, regular cleanings, planning for shows, committee meetings, training sessions, etc. on the air.
4. Say what another station is saying at the same time. If they complain, tell them you're a ventriloquist.
5. Insist on the global installation of radio parking meters. The more you stay tuned to only one station the more you have to pay.
6. Have an "Upside Down Week," where all shows would be found in a different time slot.
7. Have a "Search Week" where all shows would not be found.
8. Have a "Traffic Jam" where stations in different cities broadcast each other's traffic reports instead of their own.
9. Play the accordion: go from one watt to full power in one watt per day increments and back down again.
10. Keep all faders up and play the entire record library of the radio station and then get rid of it.
11. Keep all faders down and wait for a phone call.
12. Fill your program with nothing.
13. Empty your program of everything.
14. Give your guest the controls and put yourself at the guest spot.
15. Dissect the equipment of your radio station into its component parts: transistors, capacitors, integrated circuits, etc. and send one out to each of your listeners.
16. Go as fast as the technology you're using. Carry your words to your listeners by running.

Written in 1992–1994 and used in a section of the lecture performance "Recipes For Disaster: post-digital voice tactics" presented in 1997 at the Recycling the Future *event organized by Kunstradio in Vienna, Austria. Revised in 2004 and first published in* Christof Migone — Sound Voice Perform *(Los Angeles: Errant Bodies Press, 2005).*

can improvise sense. It can give you access to transmissions by which you can enact your own casting of what it means to radiate. Touch that dial until it touches you back.

★ ★ ★

This text[1] was the curatorial statement for an event curated by Christof Migone and Jean-François Renaud. This event comprised an exhibition, a workshop, a performance evening and a panel discussion. It took place August 8 to September 12, 1990, at the SAW Gallery, Ottawa, Ontario.

The exhibition included the installations *Sputniks* by Nicolas Collins, *L'espace voulu* by Marguerite Dehler, *There's a Mirror/Ear at the End of My Bed* by Nell Tenhaaf and Kim Sawchuk, and a video tape by Mbanna Kantako (born Dewayne Readus) entitled *One Watt of Truth* on the activities of his 1-watt pirate station WTRA in East Springfield, Illinois, and the ensuing struggles with the FCC.

In addition, the following works were available for listening by the gallery visitors: Jacki Apple and Keith Antar Mason, *Frenzy in the Night*; L'ACRIQ inc., *J'aurais pu l'écraser*; blackhumor, *no lust for the wicked*; D. Morris, *Flag Air Base*; Andrew Herman and P. Cheevers, *The Skull Bubble*; and Hildegard Westerkamp, *Kits Beach Soundwalk*.

On the evening of August 10, 1990 there were the following performances: John Oswald, *Plunderphonology: A Polystomatic Dissertation*; Gregory Whitehead, *Terror Glottis*; and Bruire (Michel F. Côté, Robert M. Lepage, Martin Tétreault), *Muss Muss Hic*!

Dan Lander conducted a workshop over three nights entitled *The Referential in Sound*.

The symposium, *Radio as Art: Issues of Creation*, Issues of Regulation, was conducted by Paul Cheevers, Chantal Dumas, Andrew Herman, David Moulden, John Oswald, Patrick Ready, Kim Sawchuk, Claude Schryer, Philip Szporer, Dot Tuer and Gregory Whitehead. Moderated by Jody Berland.

NOTES

1. Acknowledgments: Raoul Vaneigem in *The Revolution of Everyday Life* for the three projects; Sharon Gannon in "The Culture of Sleep" for the breaths; Siegried Giedion in "Mechanization Takes Command" for the dark fragments; Genie Shinkle for the mapping; Dan Lander for the edits.

ALEXIS O'HARA

Automated prayer machine, Annabelle Chvostek (left) and Anna Friz (right)

CHAPTER 13

The Art of Unstable Radio

Anna Friz

I WAS NEVER ENAMOURED WITH THE BIG RADIO SIGNALS. As a teen searching for music and alternative culture outside of the influences of family or school, I stumbled across my local (low-power) FM campus/community station. Interference and noise seeped in from all sides of the shows I tuned in to, sometimes consuming the music I strained to hear with my clock radio balanced beside me on the pillow, yet I remained a devoted listener. Years later, as a programmer at that very same station,[1] I remember the day one of the other volunteers tuned a shortwave radio in the newsroom to the Universal Time Clock. We spent the afternoon listening to the strange, relentless tick of atomic time relayed from Hawaii and Colorado over surging waves of static — a signal I have pursued, recorded and composed with over and over, ever since.

As an emerging sound and radio artist, I focused on listening to and sampling interfrequency sounds and the mingling of stations near and far. Peripatetic signals, origins unknown, dissolve as the dial rolls toward a more powerful station, but out on the perimeters of licensed radio, on the under-populated AM dial or shortwave radio, whole radio ecologies exist in the rash of static and oscillating tones. I began small forays into such in-betweens of radio as I crafted an itinerant radio practice, jettisoning the studio for the intimacy of my house or the banality of the street to temporarily occupy radio space with small unlicensed FM transmitters, their tenuous signals varying in range with every change of location. Over ten years of involvement with micro- and low-watt transmission, I now find myself working not to stabilize or increase my signal, but to highlight the volatility and unpredictability of radio territories — to manifest and play within the radiophonic environment that I inhabit. My pirate activities, then, are about both the potential in very small stations and minor signals, and embodied explorations of the phenomenology of wirelessness and the radio imaginary.

Brandon LaBelle notes that "pirate radio broadcasting contributes to a perspective on the radiophonic imagination, by not only supplying alternative content but by defining radio's borders according to an ambiguous terrain."[2] I have become particularly interested in pulling the notion of "radio" away from point-to-point casting (narrow or broad) to better perceive and experiment with this "ambiguous terrain" which exceeds the dominion of spectrum allocation, lease, licence or ownership. For the möbius-like insides and outsides of radio space are home to more than radiant signals, legal or clandestine: this is the domain of electro-magnetic waves unmodulated by programming, but nevertheless responsive to human and electrical relations. I have explored unlicensed radio activity for critical aesthetic and political intervention, and created multi-channel transmitter and receiver arrays that enable volatile, immersive, radiophonic systems. Through these performative radio/art works I consider notions of proximity and distance, interference and feedback, radiance and resonance within the much smaller and more palpable circuit of low-watt unlicensed transmission, where radio need not be limited to an apparatus for diffusion or communication, but may also become a landscape, an imaginary no-place, and a field of relationships.

Neighbourhood Infiltrations

I built my first FM transmitter at a workshop given by Bobbi Kozinuk in 1998 at the Western Front, an artist-run centre in Vancouver.[3] The schematic was adapted from Tetsuo Kogawa's design for a 2-watt FM transmitter, and I appropriated a small pair of TV "bunny ear" antennas for the job of dipole transmitter antenna.[4] In theory, pirate radio was not such a radical political proposition for me under the circumstances, living as I did in a city like Vancouver, British Columbia, where two campus/community radio stations already broadcast a panoply of political views, music and cultural expression. As a volunteer programmer, I had ample access to the airwaves and the freedom to broadcast very experimental material. In terms of garnering listeners, 1800 watts of FM on an established frequency would seem to be much more effective than a measly 2 watts heard intermittently in shifting local areas. In practice, however, such small transmitters might effectively re-materialize radio, and propose a renewed social engagement with the medium, as listeners and senders are all close by, even in face-to-face proximity with one another. Such a portable transmitter is perfectly suited for unorthodox, experimental interventions and infiltrations, with the intent of generating unique social and/or aesthetic circumstances, and for critically questioning the conventional notions of transmission that dominate the dial, persisting even in independent radio culture. In short, building my own transmitter and turning it on provided me with a first tangible sense of the untapped potential for radio outside of the need for or restrictions of the complex economic, political or technical infrastructure of a radio station.

Tetsuo Kogawa describes many of his radio activities as "radio parties," where local and translocal transmissions provide the occasion for social gatherings.[5] In my own nascent radio practice using low-watt transmitters, these smaller fields of transmission enabled unexpected social interactions. *NRRF 90.7 FM* (No Regular Radio Frequency) was a one-day radio intervention during the *Mile-End Harmony Festival* in Montréal on April 28, 2001. The weekend before, I had been one of the thousands of protesters who were tear-gassed by police during the rallies against the proposed Free Trade Agreement of the Americas (FTAA) in Québec City, where I had also made audio field recordings of the events. Back down in Montréal, I teamed up with Richard Williams to rebroadcast the raw sonic materials of protest on the street of our local neighbourhood, using a 1-watt FM transmitter and a series of radios distributed around the table and down the street from where we were narrowcasting. Our initial intention was to inject this turbulent soundscape into the relative calm of the neighbourhood, provoking political reflection and allowing the events to literally continue reverberating out into the community.

What we did not anticipate was the degree to which people attending the street festival would be eager to discuss their impressions of and/or perspectives on the protests, as many people in the area had themselves been present in Québec City and were still actively processing the experience. As people gathered around our table to trade stories, we set up a microphone on a stand at the curb, and interviewed the neighbourhood. I will emphasize that all media outlets, from mainstream to alternative, print to television, had been covering the protests in Québec, so we were hardly the only voice on the subject. Moreover, our transmission radius was no more than a city block, and we had relatively few radios tuned to emit our "programming" out on the street. Nonetheless, many people stopped to listen to the sounds of the protest from the week before, and to the stories shared by others in person on the open mic. What was remarkable was that the intimate terms of this micro-transmission enabled an unexpected circuit of social relationships on the street that day. Passers-by found an intimate — though public — forum to discuss what had been for many a troubling, emotional event, while the radio broadcast the raw sounds that had been largely missing from the official coverage of the protests.

More directly modelled on Kogawa's radio party concept were the occasional Radio Free Parkdale events that I hosted with housemates in Toronto in 2005 and 2007. We picked a loose theme or day (for instance Halloween, or May 1), invited friends over to tell stories, play music, to take part in impromptu radio plays, or just to hang out, as they pleased. Some friends in the neighbourhood stayed home and listened, some phoned in, while others arrived for the novelty but found themselves on the mic anyway, persuaded to play a character, spin a record, or to deliver an impromptu rant. I "advertised" with flyers at cafés and businesses within range a few days before the broadcast, and we sent word out through various informal online networks, but we were never entirely concerned with who or how many might be listening. The transmission to potential listeners down the block was no more important than the transmission between the people on the mic upstairs and the friends listening to the radio in the living room.

These informal, irregular radio events were characterized by their free-form sound and participatory nature. Unlike independent licensed stations for which the programming grid still serves as the central organizational paradigm, made-at-home pirate radio can be as loose as we choose. No need to change shows at the top of the hour, no need to start or finish on time; no need to conceive of "shows" at all. The micro-radio party is not about diffusion but communication — a

CLAIRE PFEIFFER

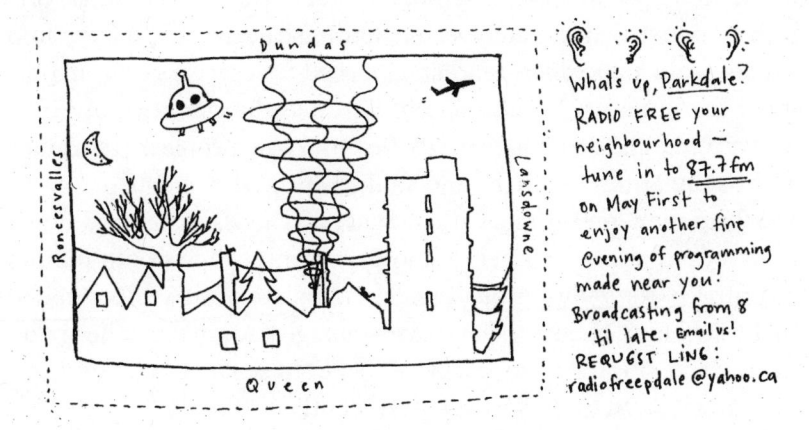

Free Radio Parkdale flyer

means for people to listen to radio together and make radio together, with transmission largely taking place face-to-face in these small transceptive circuits of interaction, adding new layers of sociality.

Who are the People in the Radio?

If such neighbourhood piracies revealed some of the potential enabled by minor media, I wondered what possibilities micro-radio poetics might yield. I wondered about the imaginary worlds proposed in even the smallest circuit of transmission — for instance, in the distance between radios in the same room — and concocted a tale of an imaginary pirate and her experience of living inside the black box itself:

> **Don't Cry Mother... It's Only a Program!**
> She shares the heartbreak of a girl who is hundreds of miles away — yes, farther than distance itself, for she lives in the land of make-believe. But it isn't make-believe to this lady because, thanks to the golden tone of her General Electric Radio, every program is close, intimate and personal — an actual visit from the interesting neighbours on the other side of the dial.[6]

When I was a child, I half-believed that there were little people inside the radio, responsible for the voices and music that came out of the radio receiver. Turn on the radio, and the little people perform, switch channels and they switch voices. But perhaps hard times have descended on these miniature denizens of Radioland, causing downsizing of the people in each radio, redundancy and finally, diaspora. Perhaps one such little radio person, finding herself alone in her radio set, not really remembering what happened to separate her from others of her kind, might take action. I created the character of Pirate Jenny from feeling a bit like one of the little radio people myself, alone in the submarine-dark halls and studios of the station where I had a free-form show, on the mic with no sense of who was listening.[7] Pirate Jenny began as an insurgent radio personality who is both *on* and (literally) *in* the radio. Alone in her set, unsure of how this came to be, Pirate Jenny is nonetheless savvy enough to transform her radio receiver into a transceiver (a device that both sends and receives), so that she can both listen for the signals of other little radio people, and send out a secret SOS when her daily duties to those she calls "the Ears" are done.

What happens when we turn off the radio? Pirate Jenny is busy monitoring the airwaves in her spare time, sending a signal intended for the long lost others of her kind. Again. Again. She is waiting for her chance to act, between the hand slapping the sleep button and the morning click of the dial that is her signal to yelp into Ear-awakening life. While the Ears are safely asleep, she ventures afar into realms of sibilance and hiss, nursing loss and threatening mutiny, searching the abandoned cities of static indicated on the old radio dials, searching for the others that she lost one day or one night, her message heard by other Ears, or neighbours, or no one, alike. Pirate Jenny hears the faint signals of other radios performing for other Ears, so she knows she is not alone. Through Pirate Jenny, radio can literally become (self) conscious. But will other little people in other radios hear her and respond? And will she be able to decipher the message if she finally receives one?

The Clandestine Transmissions of Pirate Jenny (2000-2003) was a radio art project based on this quirky fable, and took many forms: a live staged solo performance, a composition for broadcast on national public radio, a "takeover" of a community radio station and a covert pirate radio action.[8] The pirate transmissions took place in various locations in Montréal near midnight, and consisted of tuning a low-watt transmitter to a vacant spot on the dial immediately next to another much stronger station, and transmitting Pirate Jenny's SOS. In this way, anyone within range who was listening to the "legitimate" station with their radio slightly detuned, or flipping across the dial, would potentially hear Pirate Jenny's voice emerge from the static. I never knew who, if anyone, heard her pleas, but my pirate activities allowed me to closely inhabit the character and her circumstances, the experience of which I fed back into other performances of the piece. The other incarnations of Pirate Jenny on licensed radio or in live performance consisted of Pirate Jenny's SOS signal, her monologues to herself and the unknown listener, and soundscapes created from intercepted signals, radio scanning, static and noise. The actual pirate transmissions consisted only of Pirate Jenny's voice, modulated by a vocoder or walkie-talkies, but in the other iterations I needed to provide the static landscape from which her voice would emerge.

My fictional descent into the black box of radio was part of a real-life practice of radio deconstruction and remix, both as a listener and as a broadcaster. In effect, I made myself into a transceiver, listening to the city between stations, broadcasting with my homemade low-watt transmitter, invoking unseen radio territories to trouble the conventions of radio use and practice. The in-between places of the radio are not dead air zones, but uncharted airwaves rich in meaning and potential — the potential habitat of the little radio people, the mythical offspring of early radio technology. Through these soundscapes and interventions, programming and noise ceased to be binary opposites but intertwined concepts: the programming is noise, and the point is that noise is meaningful, but not representational, sound. As James Sey notes, in listening to pure frequencies, "sound is the aesthetic without representation — since there is no visual object or use of language."[9]

Pirate Jenny is constituted by the same paradoxes of immanence and imminence that characterize voices on the radio: her voice indicates her presence on the radio, but as a creature *of* the radio, where is she exactly? In my radio? In yours? In the transmission between radios? In fact, she is present within her signal range — everywhere within it and nowhere, at once. She is fictional, and yet she is "real" — audible, vibrating in the listener's ear (or Ear). She has a singular name, Pirate Jenny, and yet, when she performs for the Ears she is also all the other voices that clamour from the radio: the news anchor, the Top 40 deejay, the weather reporter, the Sunday morning evangelist and a voice, groping in the dark, to ask "who's there?" *The Clandestine Transmissions of Pirate Jenny* is a gesture toward the transceptual potential for radiophony and unlicensed activities in the air — a micro-resistance to the otherwise disenchanted radio landscape. In this fiction, the radios themselves are cast as dreamers, their uneasy sleep filled with static-y landscapes and stuttering secret texts, as they imagine the network, however small or however diffuse, of which they might be part.

The Automated Prayer Machine

Returning to the scale of early street radio interventions, what social relations are implied if the circuit of transmission shrinks to the space of a performance venue, where receivers mingle out in the audience, and the broadcast antenna is clearly visible next to electronics and instruments? Brandon LaBelle notes the close relations of transmission towers to other towers associated with political, religious, and military power or magic:

> As a formal language, the tower expresses an ongoing relation of earth and heaven, operating to channel correspondence and communications, between man and god, between church and society, and between enemies, demarcating time and space while monumentalizing historical events.[10]

What happens when transmission comes down off the tower to human scale, and seeks to express mere mortal desires? One possibility is micro-radio: modestly subversive, no longer mysterious or remote, but still able to generate something magical.

In 2004, with Annabelle Chvostek, I created a piece entitled *The Automated Prayer Machine*,[11] conceived in response to the growing sense of foreboding and despair over world events, and as an antidote to the escalation in media hysteria and increasingly sensationalistic news reporting that encourages apathy and fear rather than any kind of productive or positive action. The winter of 2003/04 seemed particularly dark in this respect, as the US-led invasion of Iraq dominated the news, while enormous international rallies for peace received relatively little attention, and did nothing to stop the actions of the Bush administration. Inspired in part by the transmission properties of a prayer wheel, where wind, water, human, or even electronically powered wheels are believed to activate a written prayer or mantra, Chvostek and I proposed a shift from apathy into empathy, and from indifference to compassion by recasting radio receivers from squawkboxes to agents of reverie. We sought to rethink radio as capable of manifesting a common space for unfolding human hopes and aspirations.

The composition of the piece relied on radiophonic sources such as live sampled radio, pre-composed samples of syndicated American talk radio,[12] and prayers that we asked friends and acquaintances to record on my telephone voicemail box. We deliberately left our definition of prayer wide open to interpretation by the people who donated prayers for us to use, so we received prayers from monotheists, polytheists, agnostics and atheists. For ourselves, we conceived of prayer as the articulation of desires, wishes or aspirations, with or without a religious context. In performance, we enhanced the radiophonic character of the piece by employing a low-watt FM transmitter to narrowcast the sound to multiple radio receivers spread throughout the audience, as well as using the standard sound system in the venue. Our circuit of transmission remained small, with the radios spread out among the audience (on or under chairs, in people's laps or on ledges), and range of broadcast occurring almost entirely within our line of sight inside the venue. Additionally, we created sound through feedback circuits within the circuit, both between the microphones and the speakers, and by sampling our signal from a radio receiver back into the narrowcast, which was then re-sampled and retransmitted. Digital video projection, acoustic instruments, such as accordion and violin, and electronics (samplers, live effects processing, etc.) completed a circuit of analogue and digital, wired and wireless practice.

As we continued to develop the Prayer Machine while on tour, we began to think more deeply about the relationship between FM radio and prayer as forms of wireless transmission. Both express something of the very mortal desire for profound communication across distance and time, while also representing the failure to realize total knowledge of or union with the Other.[13] We used prayers in many languages, particularly as we continued collecting prayers after each show while on tour across Europe, so many of the prayers were incomprehensible to some audience members. The emotional quality and even fragility of the prayers was echoed by the fragility of the radio transmission, which was prone to interference and strange bursts of static as we played in big cities where the radio dial was already completely full of licensed stations. What was left, then, of these prayers? Their power seemed located in the accumulation of many small, incidental messages of hope and good will, directed at anyone and no one at the same time; always partial, affective, but never complete, each side of the transmission unknown to one another. Faith, in other words, that it was worth speaking aloud and listening at all.

Respire

My recent performance and installation pieces work with multiple low-watt FM transmitters and an array of between 12 and 200 receivers,[14] all of which have resulted from a gradual process of introducing less rather than more stability into my interactions with radio waves.[15] For these pieces, the radio receiver array is usually suspended from the ceiling of a venue and lit only by small LED lights, which creates a dim, visually static environment. The sonic static, however, is constantly in motion.

For *Respire* I employ up to three low-power transmitters, varying in wattage from 50 milliwatts to 2 watts FM, that are effected by the spill from other existing licensed stations in the area.[16] I deliberately set the transmitters to narrowcast on related frequencies, thus encouraging multi-path and harmonic interference. As a result, the receivers emit twitters and oscillations of sound before I begin to direct any sounds through the transmitters on the various frequencies. Audience members walking among the radios may interfere with the signal from the transmitter to the receiver, causing brief bursts of sound in one or a few of the receivers, and revealing the station or interfrequency static hidden underneath. Weather, time of day, construction of the building in which the array is housed and the radiophonic environment all directly effect the sensitivity and volatility of the system, and the sounds heard. Into such a responsive radiophonic landscape, I transmit both improvised and composed sounds, favouring sounds that echo human breath, and that amplify and focus the radio environment of signal and noise. For instance, I compose with theremin and VLF (recorded very low frequencies in the electro-magnetic spectrum) signals, and sample the so-called "surplus" inadvertent sounds that bodies make on radio (intakes of breath before speaking, glottal admissions, hissing and popping air hitting the microphone, weight shifting in a chair, breath grown ragged, the overheard background from a live report, and so on), as well as live shortwave radio, walkie-talkie feedback and playing harmonica. These samples of what might be considered abject sound alternately seep or explode through the thin heterodyne music of the radios in the array, in a dynamically panned pattern that causes the sound to alternately move and hover in the space.

I describe *Respire* as a hybrid work of radio art, in that it plays both with conventions of radio content as well as with the radio waves themselves. With this piece I seek not to occupy the airwaves, however temporarily, but rather to collaborate with them, and in so doing achieve less rather than more control. The composed sounds may, at times, become almost obliterated by the sounds generated by the volatile radio environment. The resulting piece transports "noise" from the category of surplus or unwanted sound, to sound that has potential: the potential to further pry open the radio imaginary. In *Respire*, sound serves as representation, but importantly sound is also an index of the complex, changeable, embodied relationships between devices, bodies, radio waves and electricity. Pirate radio, in this iteration, is less about clandestine or subversive radiation than it is about resonance, both physical and imaginary; a realm which "resonates in our cells and allows us to share the experience with inanimate things."[17]

Coda

At a recent conference on the topic of alternate forms of exchange,[18] Greg Younging raised the question: how can something be owned which is constantly changing? Here, he was referring to indigenous peoples' perspective on creativity and culture, particularly with regard to cultural objects and customs, and proposing not ownership but custodianship over lands, language, songs, dances, symbols, etc. Younging's emphasis was on an extended sense of temporality, one that exceeds the human scale without alienating the human from it. Creative expression, in this model, is not individualistic but is held time-based, held in common, and stems from continuous, though changing, relationships with culture and landscape, through language and phenomenological experience.

Western culture, particularly under late capitalism, proposes not only that everything can be owned, but be bought and sold as well. The radio spectrum, a notion invented in the early twentieth century as a way to conceptualize an electro-magnetic territory described in frequencies (Hertz) ranging from lowest to highest, includes the more pedestrian areas of radio and television along with visible light, microwaves, and military radar. The image of the spectrum is a decidedly linear mapping that is both discursive and geographic, representing a flat earth to be divided up among government and corporate interests; a territory that periodically is auctioned off to the highest bidder, and a political as much as scientific construct designed to control "inappropriate transmissions" through frequency allocation and licensing.[19]

How would our image of the spectrum shift, what understanding might be possible, from a phenomenological experience of the same? What kind of radiophonic relationships might we sustain across time if we think of radio as more than point-to-point communications on a licensed frequency? *Listening* across the spectrum yields a radically different impression than the visual charting of it: frequencies are not discrete, but noisy as well as harmonic, overlapping one another and full of fluctuation, interference and dynamic activity. Declaring ownership over such oscillating territories perpetuates the relentless functions of colonialism that have consistently worked toward conquest rather than collaboration or custodianship. The alternative history of radio that pirates hold in common reimagines broadcast communication and the potential for community within the radio spectrum. Likewise, the use of unlicensed transmitters to explore and play with the materiality of electro-magnetic waves has the potential to reimagine radio as a medium. Here the notion of radio shifts from a radiant means of communication to a resonant though unstable environment that is immersive, palpable and affective: an ocean of sonorous and sibilant waves that is an index of relationships, both microscopic and cosmic in scale, with which humans may collaborate, but cannot claim to own.

NOTES

1. CITR Radio, the campus/community FM station broadcasting from the University of British Columbia, Vancouver. http://www.citr.ca/ (accessed March 15, 2009).

2. Brandon LaBelle, "Transmission Culture," in *Re-Inventing Radio: Aspects of Radio as Art*, ed. Heidi Grundmann et al (Frankfurt am Main, Germany: Revolver, 2008), 85.

3. See Chapter 10 in this volume.

4. Kogawa is perhaps best known for his part in the micro-radio boom in Japan in the 1980s, and currently for his micro- or mini-FM transmitter designs

which he continues to refine. Inspired by pirate and free radio activities going on in Italy and England, Kogawa began building micro-transmitters that could only broadcast a few city blocks at most, as a loophole in the highly restrictive state control of the radio dial. Kogawa's radio transmitters were entirely demonstrable and comprehensible to users, who were able to build their own and hold "radio parties" in the densely populated apartment blocks in Tokyo. Importantly, Kogawa did not imagine linking these transmitters together to create one larger signal, nor did he propose any universal communion or experience from these experiments, but rather encouraged a re-appropriation of electronic technology for diverse, unscripted and idiosyncratic social use within a very formal society. For his most recent transmitter schematics, see http://anarchy.translocal.jp/radio/micro/howtotx.html (accessed March 15, 2009). See also Tetsuo Kogawa, "Toward Polymorphous Radio," in *Radio Rethink: Art, Sound and Transmission*, ed. Daina Augaitis and Dan Lander (Banff: Walter Philips Gallery, 1994), 286-299; and Tetsuo Kogawa, "Mini-FM: Performing Microscopic Distance," in *At A Distance: Precursors to Art and Activism on the Internet*, ed. Annmarie Chandler and Norie Neumark (Cambridge, Mass: The MIT Press, 2005), 190-209.

    5. See Tetsuo Kogawa, "What is Radio Party?" http://anarchy.translocal.jp/radio/radioparty/index.html (accessed February 4, 2009).

    6. General Electric Radio ad appearing in *Life* magazine, 1940. Reproduced in *Sound States: Innovative Poetics and Acoustical Technologies*, ed. Adalaide Morris (Chapel Hill and London: University of North Carolina Press, 1997), 41.

    7. Pirate Jenny is also the name of a mutinous chamber maid in Bertolt Brecht's *Threepenny Opera*. The name, aside from being catchy, is a deliberate nod to Brecht, as he is also cited for his call for radio to be more inclusive and participatory for the public. "Transceiver," or a wireless device that can both send and receive, is a term Brecht uses when discussing the unrealized potential for radio in 1932. See Bertolt Brecht, "The Radio as an Apparatus of Communication," (1932) in *Radiotext(e)*, ed. Neil Strauss (New York: Semiotext(e), 1993), 15.

    8. *The Clandestine Transmissions of Pirate Jenny* was first performed at Ace Art gallery during *Send+Receive Festival of Sound*, Winnipeg, in October 2000. See also ORF Kunstradio's documentation of the composed piece: http://www.kunstradio.at/2002A/10_02_02.html (accessed February 4, 2009).

    9. James Sey, "Sounds Like...: the Cult of the Imaginary Wavelength," in *Radio Territories*, ed. Erik Granly Jensen and Brandon LaBelle, (Los Angeles and Copenhagen: Errant Bodies Press, 2007), 25.

    10. LaBelle, "Transmission Culture," 68.

    11. *The Automated Prayer Machine* was created as a touring piece for the HTMlles festival circulation .01, produced by Studio XX in Montréal in 2004. In 2004, Chvostek and I performed it at the *Digitales* festival in Brussels; Club Transmediale at bootlab, Berlin; a live in-studio performance on ORF Kunstradio (Austrian national public radio) http://kunstradio.at/2004A/08_02_04.html (accessed March 15, 2009); Wohlfarht, Rotterdam; Rote Flora, Hamburg; A4, Bratislava; Rad'a Gallery, Montréal; Send+Receive festival of sound in Winni-

peg; and the Western Front, Vancouver. We are grateful for the support of Canada Council for the Arts, Media and Audio Art section, for supporting our tour.

12. Thanks to Emmanuel Madan, who allowed us to sample from his substantial archive of American talk radio material, that he recorded in September 2002, while on a road trip through the U.S.A. For more information on his audio/radio project *Freedom Highway* which came from this material, see http://www.freedomhighway.org/ (accessed March 15, 2009).

13. "Dream of total union: everyone says this dream is impossible, and yet it persists." Roland Barthes, *A Lover's Discourse: Fragments*, trans. Richard Howard (New York: Hill and Wang, 1993), 228. John Durham Peters makes an eloquent case against the long-held ideal of human communication, that of profound union with others that transcends material differences and the body itself. Peters demonstrates that our longing for and ongoing failure to achieve the impossible has fundamentally shaped how media have evolved and continue to fuel communication breakdowns. "Only moderns could be facing each other and be worried about 'communicating' as if they were thousands of miles apart." His remedy is to embrace the partiality and fallibility of human intercourse, to unseat the principle of dialogue as the preferred mode of communication and recuperate dissemination, while championing a compassionate pragmatism that touches but does not presume to unite with the other. See John Durham Peters, *Speaking into the Air: A History of the Idea of Communication* (Chicago: University of Chicago Press, 2000), 2.

14. These works include: *La vida secreta de la radio/The Secret Life of Radio*, 2005; *You are far from us*, 2006-2008; *Somewhere a voice is calling*, 2007 (with Absolute Value of Noise and Glenn Gear), and *Respire*, 2008.

15. For more on contemporary art works specifically dealing with electromagnetic waves, see Armin Medosch et al, eds. *Waves: Electromagnetic Waves as Material and Medium for Arts*, Acoustic Space Lab #6, (Riga, Latvia: RIXC Centre for New Media Culture, 2006); Inke Arns, "The Realization of Radio's Unrealized Potential: Media-Archaeological Focuses in Current Artistic Practice," in *Re-Inventing Radio: Aspects of Radio as Art*, ed. Heidi Grundmann et al (Frankfurt am Main, Germany: Revolver, 2008), 471-492.

16. *Respire* premiered at RadiaLX 2008 at the Fábriça Braço de Prata in Lisbon, Portugal, September 26, 2008. The piece is closely related to another performance/installation piece *You are far from us* (2006-2008), functioning as a kind of remix of similar themes. Thanks to Conseil des arts et des lettres du Québec: Arts médiatiques for travel funding.

17. Sey, "Sounds Like...: the Cult of the Imaginary Wavelength," 22.

18. *Copyright's Counterparts: Alternative Economies of Exchange*, SSHRC Summer Institute, Queen's University, Kingston, Ontario, August 6-9, 2008.

19. See Zita Joyce, "Electromagnetic myths, ether vibrations in the space between the worlds," in *Spectropia: Illuminating Investigations in the Electromagnetic Spectrum*, ed. Daina Silina et al (Riga, Latvia: RIXC Centre for New Media Culture, 2008), 144-150.

KRISTEN ROOS

The micro radio project flyer

CHAPTER 14

Repurposed and Reassembled
Waking Up the Radio

Kristen Roos

THIS ESSAY OUTLINES MY APPROACH TO CREATING radio art, by winding back and forth between inspirational texts and recordings. I am also interested in exploring how audio technology is repurposed, whether by accident, or within the realm of sound art, hip hop and radio art. In the process, I will explore some of the influences on my Micro-Radio Project, as well as describe the places and spaces in which this project's unlicensed broadcasts have taken place since 2005.

I've downloaded the first text from the C-Theory website, and it's right here on my laptop; it's "The Turntable" by Charles Mudede. In this essay, Mudede speaks of the idea of repurposing through the use of the turntable in hip hop. I'll scroll down to a quote that I like; it's under the heading "scratch 5":

> The turntable is always wrenched out of sleep by the hand that wants to loop a break or to scratch a phrase. In a word, the turntable is awakened by the DJ who wants to make (or, closer yet, remake), music (or, closer yet, meta-music); whereas the instrument always sleeps when it is used to make real music.[1]

Let's press pause on that thought, turn away from the bright screen of my laptop, take a photocopy of a magazine off my shelf — an *Artforum* from 1972, dust it off and open it up to an essay by Robert Smithson entitled "Cultural Confinement."[2] In this article, Smithson writes on what he feels is a need for artists to create works of art that are outside of the gallery and museum systems. Here's a quote that comes part way into the article:

> Artists themselves are not confined, but their output is. Museums, like asylums and jails, have wards and cells — in other words, neutral rooms called "galleries." A work of art when placed in a gallery loses its charge, and becomes a portable object or surface disengaged from the outside world.[3]

Although Smithson isn't a radio artist, I mention his article because it represents one of the key moments in which site-specificity was defined. Artists who were a part of this newly expanded sculptural field took into account their surroundings; sculpture was often created from materials found in the very place that the work came to exist. The more recent emergence of a movement of site specific sound and radio art can, in some respects, pay homage to sculpture artists, such as Robert Smithson, Nancy Holt and Gordon Matta-Clark.[4]

How do the ideas of Mudede and Smithson relate to one another? First of all, they are both concerned with the creation of living art versus art that has lost its charge or is "asleep." They are both speaking of creating space that is alive, and of waking up the objects that are around us by giving them a new purpose, or reshaping them into a new form. If we take this view and apply it to an everyday object like a radio for example, we would see this radio as a box that is comprised of wires, speakers and electronic components, and as a tool for receiving wireless aural information. With this view in mind, the radio becomes fertile ground for the creation of something new. If we expand these ideas to a larger vision, the world becomes something that is unlimited in its potential for repurposing, rearranging and assembling. Thoughts of needing a licence to send information to this box can't exist within this expanded vision. Instead, this vision brings forth new ideas and new art forms that question the prefabricated reality that surrounds us.

The Micro-Radio Project emerged from this expansion of my reality, and my visions of the world as a potential work of assemblage. What if I ignored the pre-packaged, licensed spaces that dominate the airwaves on this medium we call radio? These thoughts led to a project that explored the potential of radio, as a tool for site specific sound art, and as an instrument in a larger composition. I started this project in 2005 with a three-part broadcast that examined public spaces in Victoria, British Columbia. Three separate broadcasts took place in a mall, a parking lot and a community commons garden. The broadcasts spoke of the aural environment that existed within these spaces, and how various members of society used these spaces. I used a small USB transmitter[5] hooked up to my laptop, capable of broadcasting 150 feet. Photocopied invitations to the broadcasts were posted around the city, and the audience was encouraged to arrive with radios and radio Walkmans, and to listen on their car radios. *The Parking Lot Broadcast* appears in the 2007 publication *Radio Territories*, along with a written description of the work.[6]

Let's make a cut here and rewind back in time once more, then go up to another shelf and take down the book *Salt Seller: The Writings of Marcel Duchamp*.[7] This book contains sketches from Duchamp's *Green Box* for a future aural sculpture — a sound painting that the audience can step into, or as Duchamp sketches, a musical sculpture. Duchamp also coined the term readymade, and viewed the world in a similar way to the example I gave earlier — as a potential work of assemblage. In order to elaborate on this idea, I'll take down another book from my shelf: Allen S. Weiss' *Phantasmic Radio*.[8] In the essay "Radio, Phantasms, Phantasmic Radio," Weiss looks at radio as a phenomenon, relating it to the process of sound entering our bodies, and how these sounds are organized according to various themes in our minds.

Weiss' ideas are an inspiration for my process in creating sound for the Micro-Radio Project; I explore the neural process of mixing and creating sound collage and microscopic sound sculptures on a daily basis, largely because sound is constantly entering and vibrating within our bodies. We then transmit and receive these collages to and from one another. This process relates to Duchamp's *Musical Sculpture*, by examining our method of taking in information and communicating as a kind of musical sculpture, which we can then use to create our own assemblage.

From Weiss' ideas we can rewind in time once more. In order to do this I'll head over to my laptop, go online and check out UBU Web to download an essay called "The Electronic Revolution," by William S. Burroughs.[9] In this essay, Burroughs suggests using portable tape recorders to set up a kind of guerilla media, where his collaged tape cut-ups would generate street happenings intended as direct social commentary. Burroughs introduces the idea of using store bought tape recorders in ways that may not have been thought of by their manufacturers, and placing them in public spaces, in effect repurposing them and waking them up.

I have a similar approach to the radios I use in my micro-radio broadcasts, which have been placed in a variety of public spaces. These radios, however, speak more of the past and present states of the media and telematic communication spaces, in addition to the ever-changing consumer market for readymade objects. I purchase them at secondhand stores (Value Village, Saint Vincent De Paul and the Salvation Army), seeing these places as repositories for consumer objects, somewhere between the home and the landfill. Gathering from these spaces is a comment on the flow of consumer items between Canada and the countries that create them. A hypothetical map of this flow would illustrate that a radio could be made in Japan in 1985 using components built in Taiwan, shipped to Canada, and sold in an electronics section of a department store. This radio could then be acquired at a second hand store between 2005 and 2009, possibly in a different city than it was first purchased, and possibly having gone through several changes of ownership. The total amount of land and sea covered by the radio is astounding (two or three continents) as well as the total amount of fossil fuels used to ship it, and the human labour used to create it. The radios that I use in my broadcasts are objects that contain this history, and are on the verge of becoming obsolete in the face of the new possibilities of telematic technologies.

Since we still have "The Electronic Revolution" open on my laptop, let's pull up another tab and go to Tetsuo Kogawa's website Polymorphous Space, and download "A Micro Radio Manifesto."[10] Kogawa's manifesto leads to an understanding of micro broadcasting that speaks of ecology and scale, through the use of pre-existing media technologies on a microscopic level. The listener must travel to these transmissions, rather than with larger more pervasive radio frequencies that only travel to the listener. In this sense, the Micro-Radio Project creates a space around a small transmission, in which the receivers of the broadcast are also active travellers to the transmission space. This creates a temporary community of listeners, and asks questions about the spaces that make up our aural urban environment.

While we're thinking of repurposing media, let's grab another magazine from my bookshelf, a photocopy of a *Scientific American* from

1905.[11] There's an article entitled "Fun with the Phonograph," in which there are suggestions on how to record your voice and make it sound like a high pitched "Tom Thumb" by altering the speed at which it is recorded. With this image in my head, I can't help but think that the first phonograph manipulation, and perhaps even record scratching, must have occurred around 1905, and remains unrecorded, and undocumented. I picture a family sitting around the living room busting out nursery rhymes, and accidentally realizing that they can wind the record back and forth. The record feature was eventually eliminated from the more affordable machines in the 1930s, to make way for their future use as players of purchased recordings. Even in its present consumer "play only" form, the phonograph is still repurposed. I have an example of such a thing, a recording of a performance from 1939, that I can play on my laptop — John Cage's *Imaginary Landscapes 1* — composed for record player, piano and bowed cymbal.

Whether it originally happened by accident, in an avant-garde music performance or in the turntable techniques of hip hop, what we can say is that the turntable is being used in a different way than was intended by its manufacturers. This has paved the way for the radio to be repurposed as well. If the turntable had never been used in such a way, perhaps the radio might have never been thought of as a malleable tool for creating art. Radios were originally marketed in the 1920s illustrating a way for families to congregate in the newly formed family room, which replaced the Victorian parlour. They were also advertised alongside personal headphones for solitary listening. This stereotypical image of radio has stayed with us, and we can still picture images of the family sitting down to listen in the living room, or father sitting down with his personal headphones in his easy chair.

With this image in mind, the writing of the German poet, playwright and theatre director Bertolt Brecht must have been a contrast. In 1929, after the radio, the phonograph and the telephone had recently become household objects in Europe and North America, Brecht wrote "The Radio as Communications Apparatus."[12] This has a similar message to the Mudede and Smithson articles, in that Brecht wants to "refunction" radio, or wake it up — to be able to speak as well as transmit. This idea was explored in the Canadian radio art project and publication *Radio Rethink*.[13] I've got a photocopy of a few of the articles from this project on my shelf that I can quote from: "A community can be created around a low-watt transmitter that is so limited in size that listeners are most likely to be producers as well."[4] This has recently become a reality with community-oriented pirate radio stations emerging on the Gulf Islands off the mainland of British Columbia, in which the listeners are also the suppliers. However, after contributing a weekly show on a Gulf Island radio station, I came to the realization that these stations don't necessarily wake up the radio, and operators can easily fall into the trap of using a computer with automated playlists of MP3s.

What is it that I mean by waking up the radio? Mediums usually end up piggybacking on each another, and in the process of going from one technological advance to the next we often dismiss the idea of a medium being used as an art form in itself. Radio is a medium that, more often than not, is simply used to broadcast commercial recordings, rather than functioning as a potential sound source, or a site specific transmitter and receiver of sound. To illustrate this point, we can go to a performance from 1951, in which John Cage wrenched the radio out of sleep with *Imaginary Landscapes 4*, composed for 12 radios. The composition involves various notes for the "players" of the 12 radios, such as volume and tone control. Each concert is a unique, contingent event, relying upon what is on the radio at the hour of the performance. Much like the "Tom Thumb" example with the phonograph, I am sure that there were many undocumented cases of radios being used in this manner, tweaking the knobs on and off, in a kind of play that is inherent to the medium. We can go to a recording of a conversation between John Cage and Morton Feldman, that's available on an internet archive,[15] in which Cage compares *Imaginary Landscapes 4* to lying down on a beach and being able to hear multiple radios on different stations simultaneously. Even if undocumented sounds existed in a similar way, Cage takes the idea to another level, using the 12 radios not simply as receivers, but also as instruments.

I first explored using radios in a way that is akin to John Cage's *Imaginary Landscapes 4* during a residency at *La Chambre Blanche* in Québec City in 2006. This time, I installed a more powerful 12-watt radio transmitter in the artist residency space, and an antenna on the roof, capable of transmitting eight kilometres. This project involved researching the history of St. Roch, which is the neighbourhood where *La Chambre Blanche* is situated. A local historian and a pastor at the St Roch Church were interviewed and recorded. Field recordings from the neighbourhood were sculpted and collaged with the interviews, and were broadcast to the neighbourhood. Once again, I distributed photocopied information on the broadcasts including the frequency, 97.5 FM. It was through dropping off information in mailboxes that I was able to meet several people who lived in the neighbourhood, visit their houses for dinner and document them tuning into the broadcast. These broadcasts evolved into a performance that took place in the gallery space that was available. The collage was orchestrated, and radios were walked into the gallery space and turned on and off at different intervals in the piece by myself and one other performer. The performance at *La Chambre Blanche* was my first attempt at using radios as instruments, or players, in a larger composition, and has created a form for future performances.

Even with Cages' radio playing techniques, the act of transmitting as an art form is a very recent phenomenon. Radio transmitters, unlike phonographs and telephones, have never been marketed for the public to use. The ability to transmit was in the hands of professionals, and for the most part has remained in the hands of professionals. Broadcasting corporations were established as soon as the radio made its way into the living room, therefore defining the parameters for what radio would become to the public. Radio transmitters have never been marketed in the same way that the phonograph was marketed for home use and playful activities. I'd argue that this is why the turntable was used as an instrument before radio was, and why radio art is in the process of catching up to the long history of turntablism. It's only now that people are speaking of radio as an instrument.

With this in mind, one of the first examples of radio being used as a space to transmit sound art into the living room is Pierre Schaeffer's work created in the studios of Radiodiffusion-Télévision Française. His earliest piece entitled *Railroad Study*, was broadcast in 1948, and involved a three-minute 78 rpm vinyl recording being broadcast at 33 rpm. The slow motion train sounds were reconstructed into a new composition that resembled an industrial factory. The same techniques written about in the 1905 *Scientific American* article, as playful fun activities, were now used by a musique concrete pioneer on a national radio forty-three years later. Radio was beginning to become an art form.

Another similar instance was Glenn Gould's *Idea of North*, which was broadcast by the CBC in 1967.[16] This was the first of what is now known as his *Solitude Trilogy*. In Gould's three-part sound collage for radio, he sculpts recordings of conversations using reel-to-reel audio tape, to create what he called *Contrapuntal Radio*. Gould's sonic techniques were mistaken for a mistuned dial by listeners, as there were often two or three voices collaged together simultaneously. This technique was groundbreaking at the time, and has become commonplace in today's world of sound art.

It's important to note that the work of Schaeffer and Gould was created within the confinements of licensed radio, at a time when experimental studios were funded alongside large radio stations. Such funding does not exist today, and the need for unlicensed pirate radio art on the airwaves is even more relevant in today's world of oversaturated airwaves catering to the Top 40 and classic rock. A prime example of the motives of publicly funded national radio is the decision to cut CBC radio's program *Outfront* in 2009. This was the only CBC-funded program that often helped furnish the tools for the creation of Canadian radio art and showcased the stories of "everyday Canadians."

As a reference to some of the early audio techniques I have mentioned, I composed a piece that was performed at the sound and media festival *Signal and Noise* in Vancouver in 2007. I had field recordings cut to lacquer records and recorded on audiotape, and mixed these sounds with the micro-pirate broadcast of field recordings from my laptop and USB transmitter capable of broadcasting 150 feet. In effect, I was able to use three different generations of sound recording technology in one performance. The composition recreates several journeys I had taken repeatedly while living on a northern Gulf Island, off the coast of British Columbia, and orchestrates these journeys through mixing field recordings. From a winter ferry crossing, to an ocean with American widgeons floating around, to a roadside with frogs chirping in the ditch and cars passing by, to a train ride down Vancouver Island — these are the sounds that are part of a rich rural soundscape, in which every sound has its own place. Inspiration for this work came from artists such as Hildegard Westerkamp, and from the history of soundscape study, electroacoustic music, acoustic ecology and soundwalks that are prevalent on the west coast of Canada.

The techniques that I am exploring in my work (collecting recordings from different environments and using them as material for sound collage on the airwaves) were established as early as the forties, and continued their use through the fifties and sixties. The 1960s and 1970s became a time in which pirate radio stations revolutionized radio programming, most notably Radio Caroline,[17] floating offshore in the Thames Estuary, and Radio Alice in Italy.[18] There are however, few examples before the 1980s in which a radio transmitter was used as an instrument by individuals outside the context of radio programming. One of the few examples we have is Max Neuhaus's *Drive In Music* in 1967, which is described in an essay in *Background Noise* by Brandon LaBelle.[19]

Taking out my laptop, I'll now open a digital copy of it on the internet:

> Situated on Lincoln Parkway, in Buffalo New York, the installation consisted of a series of seven radio transmitters located intermittently along a half mile stretch of the roadway. Each transmitter broadcast a particular frequency, thereby defining a particular area or zone of the roadway by giving it its own sound signature. Listeners could hear the work while driving down the roadway, tuning into the specific radio frequency each sound mixing and overlapping as one drove through one zone and into the next.[20]

Neuhaus's work provides inspiration for today's pirate radio artists. *Drive In Music* gives a new purpose for the radio transmitter, allowing it to become an instrument in a larger site-specific contingent composition. This combines the ideas of Smithson, Duchamp, Mudede, Cage and those in Tetsuo Kogawa's most recent work, in which he uses his hands as transmitters and receivers.[21] It is in this work that we can find inspiration for a new generation of Canadian radio artists who are reinvestigating the theremin and the Ondes Martenot[22] as radiophonic instruments; artists that are beginning to use, as Anna Friz states, "radio as an instrument."[23]

This truly is an exciting place to be, in terms of radio finally having become an art form in itself, aside from any licensing or corporate agendas. Artists are seeing the transmitter and the radio for what they really are — tools that have no predetermined purpose that can be shaped and moulded to create new audio experiences.

NOTES

1. Charles Mudede, "The Turntable," in *Life In The Wires: A CTheory Reader* (Canada: New World Perspectives/CTheory Books, 2004).

2. Jack Flam, *Robert Smithson: The Collected Writings* (Berkeley: University of California Press, 1996).

3. Paul F. Fabozzi, *Artists Critics Context* (New Jersey: Prentice Hall, 2002).

4. Artists in the late 1960s pushed the notion of sculpture, by creating directly in, and from, the place that the work came to exist. Walter De Maria, Nancy Holt, Robert Smithson, and Michael Heizer were some of the first to create this new form of site specific art, often referred to as *Earthworks*. The work of Gordon Matte Clark, in the late 1960s and early 1970s, continued with a similar approach in more urban locales. His most known works are his *Building Dissections*.

5. A description of the USB transmitter that I use can be found at: http://www.canakit.com/ (accessed September 12, 2009). The USB FM transmitter appears as a sound card for your computer and any audio produced by the computer will be transmitted over the USB connection to the transmitter for reception on any standard FM radio. Power for the transmitter is provided through the USB port and therefore there are no batteries required. The transmitter offers a choice of seven frequencies selectable through an easy to change DIP switch. The frequency is PLL (Phase-Locked-Loop) synthesized for high-quality stereo reception.

6. Here is an excerpt of the description of "The Parking Lot Broadcast" from *Radio Territories*: "The parking lot has a particular atmosphere, situated between an emergency shelter for the homeless on the right, and a derelict building (that was once a train station) on the left. The abandoned train station has a history of squatting, and currently stands with plywood over the windows and doors. Homeless people use the back and sides of the building as a place of refuge. The shelter to the left of the parking lot is another place of refuge. At all hours the homeless use the doorway as a place to wait, sit or sleep. The parking lot that stands between these two buildings has the remnants of clothes, needles and condoms; one must look closely to see these details, and the many levels of *repurposing* by various members of the community that are taking place; the broadcast therefore exists as just one of the many levels of the repurposing of space that is found in the parking lot." Kristen Roos, "The Parking Lot Broadcast," in *Radio Territories*, ed. Erik Granly Jensen and Brandon LaBelle (Los Angeles: Errant Bodies Press, 2007).

7. Michel Sanouillet, *Salt Seller: The Writings of Marcel Duchamp* (New York: Oxford University Press, 1973).

8. Allan S. Weiss, *Phantasmic Radio* (Durham, NC: Duke University Press, 1995); Allan S. Weiss, *Experimental Sound and Radio* (New York: TDR Books, 2001).

9. William Burroughs, *The Job: Interviews with William S. Burroughs* (New York: Penguin Books, 1989).

10. Tetsuo Kogawa, "Micro Radio Manifesto," *Polymorphous Space*, http://anarchy.k2.tku.ac.jp/index.html (accessed September 12, 2009).

11. "Fun with the Phonograph," *Scientific American*, 1905.

12. Marc Silberman, *Brecht on Film and Radio* (London: Methuen Press, 2000).

13. Daina Augaitis and Dan Lander, ed. *Radio Rethink: Art, Sound and Transmission* (Banff, BC: Banff Centre Press, 1994).

14. Ibid.

15. John Cage and Morton Feldman conversation available as a streamed mp3: http://www.ubu.com/sound/cage_feldman.html (accessed April, 2009).

16. Tim Page, *The Glenn Gould Reader* (New York: Knopf, 1984).

17. Radio Caroline is mentioned within a larger context of radio transmission, in Brandon Labelle's article "Transmission Culture", Heidi Grundmann et al, eds. *Re-inventing Radio: Aspects of Radio as Art* (Frankfurt Am Main: Revolver, 2008).

18. Mikkel Bolt Rasmussen, "Promises in the Air: Radio Alice and Italian Autonomia," in *Radio Territories*, ed. Erik Granly Jensen and Brandon LaBelle (Los Angeles: Errant Bodies Press, 2007).

19. Brandon LaBelle, *Background Noise* (New York: Continuum, 2006).

20. Ibid., p. 155.

21. Tetsuo Kogawa describes the use of his hands in his most recent radio performance in "Radio in the Chiasme," Heidi Grundmann et al, eds. *Re-inventing Radio: Aspects of Radio as Art* (Frankfurt Am Main: Revolver, 2008).

22. Owen Chapman explores the history of the lesser known Ondes Martenot, as another radiophonic instrument, in *Radio Activity: Articulating the Theremin, Ondes Martenot and Hammond Organ*, which is included in the radio activity volume of the online publication: Owen Chapman, "Radio Activity: Articulating the Theremin, Ondes Martenot and Hammond Organ," *Wi: Journal of Mobile Media* http://wi.hexagram.ca/ (accessed April, 2009).

23. Anna Friz writes about her use of the theremin as a radiophonic instrument in her essay "Radio As Instrument," which is included in the radio activity volume of the online publication: Anna Friz, "Radio As Instrument" *Wi: Journal of Mobile Media*, http://wi.hexagram.ca/ (accessed April, 2009).

ANDREA LALONDE

The Mysterious Death of WB, *presented by*
Small Wooden Shoe Theatre

CHAPTER 15

Radio Ballroom Halifax

Stephen Kelly and Eleanor King
(with Marian van der Zon)[1]

Marian van der Zon: What is Radio Ballroom?

Stephen King & Eleanor King: Radio Ballroom began as a pirate station based out of our home in Halifax. The programming was mainly radio art and live local independent music. The project has since grown to encompass any micro-broadcasting effort we produce. The name Radio Ballroom refers to the Ballroom Gallery at the Khyber Centre for the Arts, which originally commissioned a radio-art curatorial project in 2002. Our original intention was to broadcast from the Khyber building itself, but due to legal issues surrounding pirate broadcasting, we chose to keep the station in our home and reference the gallery space in the title.

MvdZ: Why did you start the station?

SK&EK: We initiated a few instances of FM broadcasting for live performances prior to 2002, but Radio Ballroom was our first concentrated effort at a long series with a regular weekly time slot. For this project we invited other people to interpret what it meant to use the radio itself as a medium, as opposed to simply a vehicle, for audio art or music. We were inspired by *Radio Rethink*[2] and the "KunstRadio Manifesto of Radio Art"[3] as starting points. Our aim was to invite artists that worked in a variety of other media to consider the phenomenon of radio broadcasting and create a new work specifically for Radio Ballroom.

MvdZ: Was this accomplished?

SK&EK: The artists' interpretations of the concept varied widely, but the most successful projects were those that had the energy of live performance, or utilized the radio space actively in some way. Jacob Zimmer's *The Mysterious Death of W.B.* (Small Wooden Shoe Theatre) created a three-part radio drama which harkened back to the early days of radio, complete with live actors, foley artistry and a live studio audience consisting of as many people as we could cram into our kitchen. Leah Garnett's *Live Drawing* had Leah and host Darla Kitty (Eleanor King's alias) describing exquisite corpse drawing exercises over the telephone, encouraging listeners to draw along and send their drawings back to Radio Ballroom. Andrea MacNevin created a "synchro-cast" which could only be fully heard with two radios. With cooperation from CKDU FM (the local community/campus radio station) a stereo track was played simultaneously with one half over the Radio Ballroom airwaves, and the other on CKDU's frequency. Darla Kitty's own *Live Softball* was the inaugural Radio Ballroom broadcast, a live play-by-play account of a softball game on the commons, which she announced while playing. This tongue-in-cheek parody of sports announcing was also a call for community involvement, as it invited new players to come out and join the teams.

MvdZ: How did using pirate radio inform the project, or why did you choose to use pirate radio?

SK&EK: DIY (do-it-yourself) broadcasting is like outlining an area on a map. It is site-specific and usually responds to "the local." This is evident in the history of micro-radio, where small broadcast areas ensure that programming is specifically geared towards a local listenership.

Pirate radio is also a way of temporarily claiming a tiny piece of the radio spectrum as if it were a publicly accessible resource. This act is in large part a response to the commercialization of the airwaves in Canada, where corporate radio dominates. Radio Ballroom gives us and other artists the opportunity to use radio broadcasting as a medium for art making without regard to CRTC (Canadian Radio-television Telecommunications Commission) regulations or the conventions of standard radio formats. Time limitations, mandatory weather, news and community announcements, the constant back and forth of talk-music-talk-music, and the irrational fear of "dead air" are the tropes of all official broadcasting, including commercial, public and community radio. We wanted to create a totally open broadcasting environment without being beholden to these conventions.

That being said, we still followed the "rules" to keep ourselves from getting shut down. We kept profanity to a minimum and we followed pirate etiquette by ensuring that our transmission setup was technically sound and did not unintentionally interfere with other broadcasts or radio frequency equipment. In our research we found that, unlike the FCC (Federal Communications Commission) in the US, who actively seek out and punish radio pirates, it seemed as though the CRTC and Industry Canada were mainly complaint based, so we aimed to make radio that wouldn't be unnecessarily prone to complaints. We also kept a low profile during the main Radio Ballroom season, refusing interviews even though the press was interested in the project.

MvdZ: What was the range covered your transmitter? Why did you choose this range?

SK&EK: Range is not so much a choice, but a function of the technology available to us. Our transmitter, at its best, broadcasts at about 25–35 watts (comparable to the local campus/community station at the time) and it reached most of peninsular Halifax and across the harbour to downtown Dartmouth. Examining our transmission range was one of the most magical parts of the process. We'd set up the transmitter, strap clamp the antennae on a 30-foot pole to the chimney of our house, then drive around the city to observe and take note of how far reaching our range actually was.

MvdZ: Is Radio Ballroom still active? In what ways?

SK&EK: We have not broadcast from our home in quite some time, so Radio Ballroom, as originally conceived, can not be considered active. However, we are often invited to participate in festivals, exhibitions and residencies based on our previous projects incorporating micro-radio technology. We were participants in the Deep Wireless festival in 2007 with a residency and installation called *Radioroam* which uses a low-powered transmitter as the central technology to the work. Last summer we taught a two-day workshop at the Banff Centre, where participants built simple FM transmitters, then recorded and edited localized field recordings, which culminated in an outdoor collaborative installation. We were recently part of a group exhibition at NSCAD University, and there are projects coming up in the next two years that incorporate radio transmitters.

NOTES

1. This e-mail interview was conducted by Marian van der Zon with the founders of Radio Ballroom, Eleanor King and Stephen Kelly.
2. Daina Augaitis and Dan Lander, Eds. *Radio Rethink: Art Sound Transmission* (Banff, British Columbia: Walter Phillips Gallery, The Banff Centre for the Arts, 1994).
3. Kunst Radio, *Toward a Definition of Radio Art*, http://www.kunstradio.at/TEXTS/manifesto.html (accessed September 11, 2009).

DESTANNE LUNDQUIST

CHAPTER 16

The Power of Small
Integrating Low-Power Radio and Sound Art

Kathy Kennedy

HAVING TRAINED AS A CLASSICAL SINGER, I WOULD never have expected to become a radio pirate 20 years later. My work has been about the voice and its interface with technology through music-based performance. During my career, I've seen the onset of the web and the development of network practices. In the 1990s I was involved in the founding of Studio XX, a digital media centre for women in Montréal. I've also been involved in electroacoustic music, but what I value most is my work as a community artist. Low-watt pirate radio has been central to all this work. The level of freedom and agility it gives to all my performances has been critical for getting the work accomplished. I also have to admit that I've always wanted to be a pirate. Doesn't everyone want to be a pirate sometimes?

Low-watt radio has been part of my art practice for many years, and I would like to outline my evolution in working with this medium. I've been less concerned with the idea of a radio station than with using unlicensed radio as an integral device for site-specific performances. Travelling with my personal arsenal of transmitters has proven very efficient for creating these installations because piracy allows for portability, and I can easily pick up new radio receivers at nearby thrift shops. The radio acts as a medium for content, a logistical tool for coordination, and also serves as an important metaphor for the body. Imagine a swarm of women holding radios as they stroll through the city!

201

How I came to work with this medium is somewhat serendipitous. At the Newfoundland *Sound Symposium* in 1991, I had become intrigued enough with radio to participate in a sonic joy ride with such legendary audio folk as Christof Migone,[1] Dan Lander and Claude Schryer. They may deny the accusation, but I clearly remember driving aimlessly around St. John's with a battery-powered transmitter and microphone. We wailed and hooted across the open airwaves, living up to the glorified image of pirates, or perhaps of out-of-control teenagers. As an action, it may have had only a small impact on the few listeners who had accidentally found us while tuning their radio dial. But for me it opened a realm of performance possibility that still feels new and exciting.

One of my first radio compositions involved the strategic placement of radio receivers, their manipulation by live performers (who also sang), and a musical soundtrack broadcasting from a personal transmitter. Until that point, most of my performance work had been dedicated to solo and choral pieces. I had always, however, felt limited by the concert hall and longed for a device that would allow my music to become more immersive and participatory. The inherent freedom of low-watt radio made it the perfect tool for the job. I still have not seen or heard of anyone else using radio in this particular way, although the boom box orchestra of Phil Kline in New York has been named in comparison, but those works use tape decks that cannot be completely synchronized.

In this era where high technology is unequivocally praised, radio serves as a refreshingly easy and inexpensive alternative. When my colleagues in electroacoustics were using state-of-the-art eight-channel diffusion systems, I prided myself on using a "poor person's" version with moveable humans and portable radios. In all of these works, a soundtrack is broadcast through my personal transmitter to any number of portable radios, strategically placed for live performers to interact with. My colleagues have been developing increasingly sophisticated methods to move sound around a room through complex speaker pans and arrays. I was lucky enough to skip that step and go straight to having live performers move the speakers around the room. This device has served to spread live sound around a large or complex acoustic space, filling it with a powerful auditory presence, and yet with no real loudness in any one place. The radio receivers can also act as sonic placeholders to mark a terrain for performers,

KATHRYN WALTER

Kathy Kennedy leading a soundwalk in Banff during
PRIVATE INVESTIGATORS

drawing boundaries by the range of audibility created by these small satellites. Because radio figures prominently in the discourse of acoustic ecology, a discipline that is central to my art practice, I began to wonder about the integration of radio with nature. This led to a new series of performances.

Radio Soundwalks

At the Banff Centre in 1993, I started creating radio-based soundwalks. The term "soundwalk" refers to the acoustic ecology practice of walking along a designated path and listening as deeply as possible, for between 30 minutes and one hour. It's a pleasant activity for a group, particularly for those of us who are bored to tears of recitals. However, if even one person in the group is not interested in listening, then the exercise becomes futile for all involved. The soundwalk becomes a painful game of shushing the offenders who seem to only want to do it more when they realize they are breaking the rules!

By integrating radio into performance, I found a kind of tool that could act as a common thread, to link beginning to end and to contextualize the whole. I was trying to heighten the listener's experience of exploring sound in space. The idea of using radio was to connect all listener/spectators through the transmitted soundtrack that was just barely audible as they walked from one radio-based site to another. In the soundwalks, there are musicians stationed at each radio site who improvise with the soundtrack at a volume just barely loud enough to be heard at the very next site in either direction. The sounds are therefore only loud enough to be heard one site away, and not at the third or fourth site, which creates a very specific kind of audio experience — one that relies completely on the radio to carry the sonic adhesion. Musicians and spectators at the performance are constantly aware of the physical range of their sound level. This attention to low volume throughout creates a dreamlike intimacy, a relationship of sound with physical space. It is a process of discovering our physical environment through sound transmission.

The first radio soundwalk was outdoors in a patch of forest no more than one kilometre long. Five radios were spaced equally apart with five improvising musicians playing near each of them. The soundtrack had a recording of my voice singing softly, extended tones and breaths and also reciting a quote from John Cage about silence (the fact that there is no real silence, and that even in an anechoic chamber we can still hear our heartbeat or other bodily sounds). Many would say that the sounds of nature are best left alone, but I felt strongly about adding an emblem of technology into the physical environment. To lead the listener into appreciating a pristine natural environment is not, in my view, an artwork, but plain common sense. I was trying to incorporate technology with the lightest, most personal touch possible, leaving an artifact of my own voice in the forest.

The second radio soundwalk was based on the sounds of pianos. It took place along a row of practice modules, each containing a piano and a radio playing a piano-based soundtrack. A pianist in each room was improvising with the soundtrack, and with the other pianists if they could be heard. Each of the practice room doors were opened and closed so that listeners could have a private listening experience if they liked, poking in and out of any room in any sequence. This was one of the best examples of how radio was used to unite all performers in temporality while allowing each their individual sonic space.

The third radio soundwalk took place in the photography department at the Banff Centre, using a row of dark rooms with an accompanying radio broadcast of a photography soundtrack, composed of the sounds of photographic tools. This provided an excellent opportunity for photographers to make musical improvisations over the soundtrack by using the familiar objects of their habitual work environment. The photographers made excellent improvisers, for non-musicians, slamming away at their developers, clicking knobs with great acuity and waving photographic paper in rhythmical flourishes.

The biggest radio soundwalk was at the SoundCulture '96 festival in San Francisco, through the promenade of an outdoor shopping mall. With nearly 20 improvisers creating a long passage of subtle sounds over my vocal soundtrack, the commercialized market space became a haven for listening. Passersby became influenced by the careful scrutiny of the audience, paying increasingly more attention to their own residual noises as part of the larger soundscape. Gradually, the general sound level of that space became as quiet as a concert hall, leaving an eerie silence across the mall. Locals told me later that it reminded them of the atypical quiet that happens just before an earthquake. Soon after that, this soundwalk was presented using pirate radio and called *The Blue Pathway* at the 1997 *National Campus-Community Radio Conference* in Edmonton.

With these performance pieces, I came to learn the surprising power of small sounds, and the array of effects that they create. In my work, the sound level determines the circumference of the performance, outlining the borders by the limits of audition. Everything is on a human scale, and the sound is only transmitting at a volume like what we transmit with our own bodies. When the sound is "small" (or more precisely quiet), it poses no threat; it doesn't impose. It is literally low-power, integrating with other existing sounds, rather than conquering them. The audience has the opportunity to enter the piece in a state of "reduced listening"[2] strolling, at their own pace, through the sonic environment. Each participant creates their own personal mix.

Sonic Choreographies: Neveralways, Counting Games, Taking Steps and Paradio

The idea of covering a really large physical space with sound has always appealed to me, rather like natural phenomena such as thunder or windstorms. There is nothing as exciting as the sound of many voices together. The benefits of singing, both physical and mental, can not be underestimated since the vibration of sound throughout the body has an energizing and healing effect. The "singer's formant"[3] has been widely documented for its remarkable power of volume (this is how one singer can be heard over an entire orchestra). For large-scale choral works with accompanying soundtrack, no technology compares to radio transmission for even sound distribution and personal monitoring for each and every singer. Three performances that explored this were *Neveralways*, *The Counting Game* and *Taking Steps*. They all included the powerful image of many singers holding radios, which I see as icons for the human scale of sound transmission. Each singer was part of the group but had an individual control over his or her own radio and each performer adopted the practice and, essentially, sang along.

Neveralways was my first public composition involving an outdoor, urban audience. This piece was commissioned by the *Festival International de la Voix* (1993) and performed on the grounds of Place des Arts in Montréal. The composition used the spatial placement of singers throughout the entire city block. Over 100 singers were linked by radio, singing over the recorded soundtrack, a kind of karaoke of sorts. I created a fairly elaborate soundtrack with symphonic instruments, some beats, and some *musique actuelle*. The score included a map of the city block indicating where each section (soprano, alto, tenor and bass) needed to be for each movement in the piece. It was a truly spectacular sonic occurrence — as masses of sound dispersed and converged simultaneously, wrapping themselves in and around the audience, immersing us in the experience.

The next piece expanded on these concepts, allowing me to come to know a new city. *The Counting Game* was commissioned by Western Front Gallery to mark the inauguration of the Vancouver Public Library, designed by Moishe Safdie, in 1995. The idea was to create music for choir that would favour the spatial properties of the new building. Archways, stairs, foyers and esplanades all make for different acoustic environments. I created five symphonic sections of original music to make a 20-minute piece arranged for a choir of 100 and radio transmission. The musical accompaniment for the choir was played on a low-watt FM radio transmitter (compliments of Bobbi Kozinuk), so that the choir could move freely from one end of the building site to the other. The piece was so popular that our ad hoc choir gave encore performances at the Vancouver Art Gallery and at Granville Island. The purportedly "public" space of city squares and parks is, in fact, quite restrictive of individual citizens' self-expression. Singing in public is generally associated with either madness or political propaganda. In this case, the public disturbance was fuelled by no other cause than a love of music.

The choral-radio work continued to evolve in my mind, and the next piece was decidedly in relationship with the concert hall. *Taking Steps*, was performed during the Radio Days festival in Montréal in 1995. The piece's principle element was a radio documentary on hate mobs that attack marginalized individuals, and has a very menacing tone. The sounds are of echo-filled subways and erratic footsteps. While the audience was seated in a concert hall, listening to the two speakers in the front of the room, the soundtrack was simultaneously broadcast to a room one floor above. On a very precise musical cue, a choir (wearing heavy workboots) began to stomp in a rhythmic pattern, directly above the back row of the audience just below. With each musical phrase, the stompers encroached slowly toward the front of the audience, making a threatening level of noise overhead. Perfectly in sync with the soundtrack, but on a different floor of the building, their presence was truly ominous and disturbing as if in a real mob situation, steadily and inevitably encroaching.

Paradio, the last in this series, represented the encapsulation of the concept of large-scale choral works and was performed along Banff Avenue in 1996. It was produced while I was in a performance residency called Private Investigators, and thirty or more of the residents at The Banff Centre proudly took part in this twenty-minute piece with easy vocal lines over another recorded soundtrack. Like its predecessors, it created a very marked effect on a non-artist public with very low technical requirement.

Games and Treasure Hunts

Revisiting the idea of radio-enhanced soundwalk, my compositions have recently evolved into treasure hunts, where the public is given extra motivation to listen. Commissioned by Sound Travels in 2003, the *Singing Maze* was a sonic treasure hunt that took place on Toronto's Centre Island. It is a whimsical fairy tale in which riddles are offered at various sites around the island. The story of a girl who metamorphosed into a bird unfolds, as the riddles point to fantastical sights and sounds of birds, musicians playing and singing. Using low-watt radio devices to create sonic illusions, participants are gradually led into a shrubbery maze where mysterious voices colour their path. In the *Singing Maze*, participants and children of all ages tried to find the solution to each riddle through acute listening. This piece was created as a response to the gruesome murder of a young girl, Holly Jones, in Toronto in 2003, which created a pandemic fear for children's' safety. The *Singing Maze* was intended to provide a secure context in which to wander, hide and discover.

From the top of the bridge in Centre Island, just next to a fountain is a square of lawn with 16 trees placed symmetrically. Despite their formal order, they hold a promise of the wild. Faeries (island locals in costume) could be seen running, hiding and running again in a seemingly silent dance, just in front of the *Singing Maze*. At the centre of the maze is the tallest tree. In that tree, a low-watt radio transmitter was placed, transmitting out to a dozen or more small receivers hidden in the hedges. "Go this way, turn that way" were the whispered and sung commands coming out of all receivers quietly. The directions, some of which were intentionally misleading, were interspersed with lilting melodies that transformed into bird sounds. It was as if the hedges in the maze were alive with birdlike creatures, both leading and misleading them. Those who made it to the centre found me perched in the tree, singing into a tiny mike plugged into my old travel companion, the transmitter.

The Gabriola Sound Treasure Hunt (2007) was another treasure hunt based on radio transmission. A 30-minute soundtrack was created based on hundreds of islanders telling about their favourite sounds on the island, that was broadcast through the village. A zine was published with riddles to find the hidden sources of audio trompes l'oreille, namely an invisible stream (portable radios hidden in a ditch playing a transmission of river sounds) and the breath of the forest turning into raven's wings (transmission composed of my breath sounds called *Soupire*). As can be seen from these examples, radio has helped me to create poetic and romantic imagery.

New Directions through HMMM

My current obsession is an ongoing performance project called *HMMM*, and the role of radio remains important. Participants hum on extended tones over a soundtrack composed of the same thing — long

KATHY KENNEDY

HMMM/Soundwalk
Kathy Kennedy 2005

X=radio, O=human

```
X     X     X     X     X     X
   O     O     O     O     O     O
               -----------
```

Radios are tuned to receive humming signal to a volume that is not to exceed a level that is heard by one performer on either side. Performers are spaced equally between radios. They hum on the same tone, different tones, or improvise other sounds and words, not to exceed a level that is heard by one performer on either side.

> The body literally vibrates with sound. (sic) This gives humans a striking advantage in relating to the environment. The act of singing permits us to open a dialogue with space so that we become flooded by its vibrations and merge with it acoustically speaking. As soon as you sing you get feedback from your surroundings.
>
> Dr. A. A. Tomatis

Score from Hmmm/Soundwalk

held tones of humming. The effect is a river of human vocal sound, soothing, organic and nonverbal. Ahhh. Participants have reported a sense of echolocation as they interpret the environment around them by hearing their own sounds in context. This calls to my mind the visceral and sensual nature of singing. So far, these pieces have been performed on my personal transmitters in Montréal, Toronto, Vancouver, Kitchener and Kingston. One particularly successful *HMMM* was in Lima, Peru in one of the noisiest parts of the city. Unable to see my personal, unlicensed transmitter, the police decided to try to arrest us for disturbing the peace. They were clearly disgruntled by having to agree that we were, in fact, making that area quieter and more peaceful. Better than trying to make people be quiet.

The metaphor of the body is unavoidable in my radio works. The body is an essential element in my work, and the radio is generally used to remind the audience of the natural range and quality of sound diffusion. The radio expands the span of audio transmission only to a degree that reminds us of the natural limitations of physical space. There is usually a kind of magic and wonder around the fact that certain bodies in the piece are, in fact connected to other bodies some distance away. Because I use many small sound sources, as opposed to a central one, my use of radios is also strategically intended for individual bodies, to be controlled individually. The radio is generally used as an extension of the body, a bridge from one body to another.

HMMM encapsulates my relationship with radio in general and, moreover, with pirate radio; in particular, how fragile and physical it is. It is a forbidden act, but we all know that it's intrinsically right. Like a voice, it is ultimately individualistic and inevitably subject to suppression. It might seem weak in comparison to newer technologies like the internet, but uniting many small and individual audio sources can create remarkably strong and powerful phenomena.

NOTES

1. See Chapter 12 in this volume for more about Christof Migone's work.
2. Michel Chion, *L'audio-vision*, (Paris: Nathan-Université, 1991).
3. "The singer's formant produced uniquely by resonance, with no supplementary effort... increases the amplitude of the voice, making it possible for a singer's voice to he heard over a large orchestra," in Jean-Francois Augoyard and Henry Torgue, *Sonic Experience*, (Montréal: McGill–Queens University Press, 2005), 108.

JESSE GERTES

CHAPTER 17

Voices In a Public Place

A Docudrama in Seven Acts
on/for Micro-radio in Canada

Roger Farr

Note on the Text

"Voices in a Public Place" is divided into seven "acts" — in the fullest sense of the word — which are titled after the seven components of a basic micro-transmitter.[1] In its current appearance as a script, the piece is intended to work as a kind of documentary poem (hence the use of citations). As text intended for performance, however, all names and citations are to be dropped in favour of the pirate techniques of sampling and appropriation.

Notes on the Speakers

Berland, Jody. Professor at York University; author of "Radio Space and Industrial Time: The Case of Music Formats," "Locating Listening: Technological Space, Popular Music, Canadian Mediation" and "Radio, the State and Sound Government," among numerous other works on culture, communications and nation-space.

Breaker, Shane. Media activist and member of the Siksika (Blackfoot) Nation; coordinator of Siksika Media, a multimedia outlet for the Siksika Nation and surrounding communities.

Brecht, Bertolt (1898–1956). German avant-garde poet, playwright and theatre director. His essay "The Radio as an Apparatus of Communication" is regarded as a precursor to interactive media and communications theory.

CRTC. The Canadian Radio–television and Telecommunications Commission, founded in 1968 in order to fortify the nation's cultural, social and economic structures through the regulation of its broadcasting industry.

Deleuze, Gilles (1925–1995). French post-structuralist philosopher; author, with Félix Guattari, of *A Thousand Plateaus and Anti-Oedipus*, as well as numerous other works on literature, philosophy, aesthetics, language and film.

Fairchild, Charles. Author of *Pop Idols and Pirates* and *Community Radio and Public Culture*.

Goebbels, Joseph (1897–1945). Minister of Propaganda in Nazi Germany from 1933 to 1945, during which time he sought to fortify the nation's cultural, social and economic structures through the regulation of its media. Goebbels was a strong proponent of a centralized, national broadcasting apparatus, and was particularly interested in the hegemonic possibilities of state-controlled radio.

Guattari, Félix (1930–1992). French media activist, philosopher and psychoanalyst; author of numerous works, several written in collaboration with Deleuze; active in the popular radio movements in Italy and France in the 70s and 80s, especially Radio Alice.

Kantako, Mbanna. Micro-radio activist with Tenants Rights Association, Black Liberation Radio, African Liberation Radio and now Human Rights Radio; considered by many to be a pioneer of the micro-radio movement.

Kogawa, Tetsuo. Micro-media activist, artist and philosopher; Professor of Communication Studies at Tokyo Keizai University's Department of Communications; author of over 30 books on media culture, film, the city and micro politics; introduced the free-radio movement in Japan; in the 1980s conducted a series of important interviews with Guattari.

Kroker, Arthur. Canada Research Chair in Technology, Culture and Theory at the University of Victoria; author of many works, including *Technology and the Canadian Mind: Innis/McLuhan*.

Marx, Karl (1818–1883). German philosopher, economist, sociologist and revolutionary; author of *Capital*, *The Communist Manifesto*, *The German Ideology*, *The Grundrisse*, *The Poverty of Philosophy* and many other works.

Slim, Mutebutton. Deejay of "Kinda The Blues" on Tree Frog Radio on the West Coast of Canada.

Sakolsky, Ron. Writer, musicologist and Surrealist poet; author and editor of numerous works on anarchism, radio and culture, including *Swift Winds*, *Surrealist Subversions*, and *Seizing the Airwaves: A Free Radio Handbook*; active with Tree Frog Radio on the West Coast of Canada.

Schreiner, Tom. Activist with numerous free radio efforts in the US and Mexico, including Free Radio Santa Cruz, Radio Watson and Zapatista stations in Chiapas.

van der Zon, Marian. Media activist, musician, writer and sound artist; founder of Temporary Autonomous Radio.

Vipond, Mary. Professor Emeritus at Concordia University; author of *Listening In: The First Decade of Canadian Broadcasting, 1922-1932*, among other works on Canadian cultural and media history.

Prologue

FX: *10-second MONTAGE of various FX from the script brought up slowly*

CHILD 1: This is not a message.

CHILD 2: It's an event.

CHILD 3: It's semiological delinquency without ideological transmission.

FX: *Cut MONTAGE at "transmission"*

CHILD 4: It's about friendship.

Act 1
RESISTORS

VOICE: Any receiver can become a transmitter, with a few minor modifications.

FX: *Someone DIALING AN OLD PHONE, followed by RINGING, fade under for the duration of the Act*

KOGAWA: Our microscopic space is under technological control and surveillance. Our potentially diverse, multiple and polymorphous space is almost homogenized into a mass. Therefore we need a permanent effort to deconstruct this situation.[2]

FX: *5 seconds of a PLANE TAKING OFF*

OPERATOR: *(As through phone)* I have a call from a lady at Bank of New York that states that the World Trade Center —

CRO: We got that already.

OPERATOR: She states on the northwest side there's a woman hanging from — an unidentified person hanging from the top of the building.

CRO: Uh-huh.

OPERATOR: Okay. That's all the information. One World Trade Center.

CRO: All right. We have quite a few calls.[3]

FX: *Announcement of Greenwich Mean Time, with short and long tones*

FX: *5 seconds of VOICES IN A PUBLIC PLACE, fade under*

DELEUZE: We've got to hijack speech . . .

FX: *5 seconds of a NAIL GUN*

DELEUZE: Something different from creating . . .

FX: *5 seconds of FACTORY NOISE*

DELEUZE: Vacuoles of non-communication . . .

FX: *5 seconds of BREAKING GLASS*

DELEUZE: Circuit breakers . . .

FX: *5 seconds of a CAMERA SHUTTER*

DELEUZE: So we can elude control.[4]

FX: *VOICES IN A PUBLIC PLACE brought up*

CANADA: Broadcasting means any transmission of programs, whether or not encrypted, by radio waves or other means of telecommunication for reception by the public . . .

FX: *ELECTROMAGNETIC INTERFERENCE*

CANADA: . . . by means of broadcasting receiving apparatus, but does not include any such transmission of programs that is made solely for performance or display in a public place.[5]

SCHREINER: *(As through phone)* I'm not interested in making micro-radio legal. I'm interested in radio as an instrument of struggle.[6]

FX: *10 seconds VOICES IN A PUBLIC PLACE full*

CRTC 1: *(Distant)* Note that the legislators did not provide a definition for "a public place."[7]

FX: *VOICES IN A PUBLIC PLACE cut*

Act 2
CAPACITORS

FX: *RADIO STATIC and dial surfing, pausing on CBC for 5 seconds before fading under for duration of the Act*

GOEBBELS: It would not have been possible for us to take power or to use it in the ways we have without the radio.

FX: *RADIO STATIC and dial surfing brought up for 3 seconds, then faded under*

KROKER: Canada is and always has been the most modern of the new world societies because of the character of its colonialism; of its domination of the land by technologies of communication; and of its imposition of an abstract nation upon a divergent population by a technical polity.[8]

FX: *RADIO STATIC and dial surfing brought up for three seconds, then faded under*

FAIRCHILD: The foundations of the current policy regime were laid with the inception of the Canadian Broadcasting Corporation's "Accelerated Coverage Plan" (ACP) which aimed to provide direct CBC services via satellite to any community with more than 500 residents.

FX: *LOON CALL*

FAIRCHILD: The CBC's pursuit of its coverage policies began with the creation of a Northern Service in 1958 and continued as it implemented the ACP in 1973; both efforts were designed to enhance official government policies aimed at assimilating the aboriginal population into mainstream Canadian society.[9]

FX: *5 seconds of a BALLOON being inflated*

GOEBBELS: Above all it is necessary to clearly centralize all radio activities, to place spiritual tasks ahead of technical ones, to introduce the leadership principle, to provide a clear worldview, and to present this worldview in flexible ways.[10]

FX: *RADIO STATIC and dial surfing brought up*

BERLAND: It wouldn't be Canada without radio.[11]

FX: *LOON CALL*

CRTC 1: Note that the legislators did not provide a definition for "a public place."

FX: *BALLOON bursts*

Act 3
COIL

FX: *THUNDER and RAIN*

CBC HOST: Now to local weather.

THE WEATHERMAN: *(With typical "broadcaster voice")* Today: A few towers sabotaged in the morning, followed by a 90 percent chance of communication late in the day. Overnight: Poetry. Friday: Occupations in the afternoon, then protests. Saturday: Protests with periods of communication.

Sunday: Slogans becoming poetry late in the day. Monday: Occupations.

Micro-radio transmissions will continue to keep spirits high over the weekend as more waves of turbulence move through the central region, ending a period of nationalized consensus. Communication is expected to last well into next week.

FX: *ELECTRO-MAGNETIC INTERFERENCE*

FX: *THUNDER*

Act 4
TRIMMER CAPACITOR

FX: *5 seconds RADIO STATIC and dial surfing*

GUATTARI: (Slowly) The evolution of the means of mass communication seems to be going in two directions.

FX: *3 seconds of intro music to CBC's "Canada Live," fade under*

GUATTARI: Toward hyper-concentrated systems controlled by the apparatus of state, of monopolies, of big political machines with the aim of shaping opinion and of adapting attitudes and unconscious schemas of the population to dominant norms ...

FX: *3 seconds of NAZI BROADCAST brought up, then fade out*

GUATTARI: And toward miniaturized systems that create the possibility of collective appropriation of the media that provide real means of communication, not only to the "great masses," but also to minorities, to marginalized and deviant groups of all kinds.[12]

FX: *Distant THUNDER*

CRTC 1: Note that the legislators did not provide a definition for "a public place."

BREAKER: Aboriginal radio stations in many Canadian communities are pirate broadcasters with no government licensing. Usually a new radio station would apply for a government licence to broadcast publicly through the Canadian Radio-television and Telecommunications Commission. The majority of Aboriginal radio stations don't follow this bureaucracy because the community doesn't believe a licence to broadcast their culture is necessary.[13]

FX: *Near THUNDER*

BERLAND: It wouldn't be Canada without radio.

FX: *5 seconds of RINGING PHONE full then cut*

Act 5
COPPER-PLATED BOARD

CBC HOST: Caller, are you there?

CALLER: Vertical integration.

CBC HOST: What did you say?

CALLER: I'm here.

CBC HOST: Uh, no — I'm here, you're there.

CALLER: Where are you?

CBC HOST: In a comfortable, mediatized space — somewhere between my desires and the state. From here I engage in rational, transparent dialogue with other Canadians about the things that most concern us.

FX: *ECHO "us" and fade out*

FX: *LOON CALL*

CALLER: Describe it.

CBC HOST: It's very modern and spacious — mostly white, with some tastefully integrated colour here and there.

FX: *ECHO "there" and fade out*

FX: *LOON CALL*

CALLER: The Pubic Sphere.

FX: *BALLOON being inflated*

CBC HOST: Correct.

CRTC 1: Note that the legislators did not provide a definition for "a public place."

CALLER: Does this mean I am outside the Sphere?

FX: *LOON CALL*

HOST: There is nothing "outside the Sphere," except anarchy and illiteracy.

FX: *BALLOON bursts*

Act 6
AUDIO CABLE

FX: *VOICES IN PUBLIC PLACE, fade under for duration of the Act*

CRTC 2: The Commission, pursuant to subsection 9(4) of the Broadcasting Act, by this order, exempts from the requirements of Part 1 of the Act and any regulations, those persons carrying on broadcasting undertakings of the class defined by the following criteria.

FX: *CASH REGISTER*

CRTC 2: The purpose of these radio programming undertakings is to allow those such as real estate agents...

FX: *CASH REGISTER*

CRTC 2: store owners...

FX: *CASH REGISTER*

CRTC 2: and local authorities to communicate to the public messages of an informative, sometimes commercial nature regarding their activities by means of ultra low-power transmitters, e.g. "talking signs."[14]

FX: *CASH REGISTER*

FX: *VOICES IN A PUBLIC PLACE brought up*

MARX: If commodities could speak, they would say: "Our use-value may interest human beings; but it is not an attribute of us, as things . . ."

FX: *CASH REGISTER*

MARX: "What is our attribute, as things, is our value. Our own interrelations as commodities proves it. We are related to one another only as exchange values."[15]

FX: *CASH REGISTER*

BRECHT: (As through a phone) Radio is one-sided when it should be two. It is purely an apparatus for distribution, for mere sharing out. So here is a positive suggestion: change this apparatus over from distribution to communication . . .

FX: *VOICES IN PUBLIC PLACE, brought up*

BRECHT: The radio would be the finest possible communication apparatus in public life, a vast network of pipes. That is to say, it would be if it knew how to receive as well as to transmit, how to let the listener speak as well as hear, how to bring him into a relationship instead of isolating him. On this principle, the radio should step out of the supply business and organize its listeners as suppliers. Any attempt by the radio to give a truly public character to Public occasions is a step in the right direction.

FX: *VOICES IN PUBLIC PLACE, brought up to full*

CRTC 1: Note that the legislators did not provide a definition for "a public place."

FX: *CASH REGISTER faint and distant*

CRTC 3: Industry Canada has no plan to exempt 5-watt FM broadcasting transmitters from authorization requirements, as such an exemption would open the FM broadcasting band to all users.[16]

FX: *VOICES IN PUBLIC PLACE fade out*

Act 7
9-VOLT BATTERY AND SNAP CONNECTOR

FX: *All FX repeated and layered to create MONTAGE over the course of the Act*

VOICE: Power is nothing without the means of broadcasting itself.

FX: *THUNDER*

VAN DER ZON: Micro-radio works to link people together. Radio becomes a space (both the studio and the airwaves) where the line between those who make radio and those who consume radio is blurred...

FX: *LOON CALL*

VAN DER ZON: Diverse ideas, cultures, experiences and politics are shared within the group producing radio as well as with the audiences in front of their radios. Because micro-radio usually has a weak signal, the purpose changes from broadcasting to narrowcasting. Aside from those listening, it inspires inexperienced individuals to get involved and promotes a sense of community or action.[17]

FX: *BREAKING GLASS*

GUATTARI: Languages of desire invent new means and tend to lead straight to action; they begin by "touching," by provoking laughter, by moving people, and then they make people want to "move out," towards those who speak and toward those stakes of concern to them.[18]

FX: *RINGING PHONE*

KANTAKO: When I talk to people, they tell me that they feel us inside of them. It's not just that they listen to us on the radio. They feel us inside.[19]

FX: *VOICES IN A PUBLIC PLACE*

MUTEBUTTON SLIM: i enjoy doing the show because it's cheaper than my prescription to prozac plus i get a chance to say unkind things about steven harper and bc ferries.[20]

FX: *BALLOON being inflated*

VIPOND: In early December, 1930, Edmonton part-time inspector W.G. Allen reported hearing on an unauthorized wavelength a voice speaking in broken English declaring that he was Comrade Trotsky on the air from Leningrad in Moscow. The following week the man was heard again, this time claiming that there were 14 Communist radio stations such as his in Canada and predicting that the workers were preparing to seize all "pianos, automobiles, radio sets, and luxuries."[21]

FX: *NAIL GUN*

CRTC 1: Note that the legislators did not provide a definition for "a public place."

FX: *PLANE TAKING OFF*

FX: *Announcement of Greenwich Mean Time*

SAKOLSKY: We want to communicate with our neighbours without a licence.[22]

X: *10 seconds of MONTAGE and cut*

NOTES

1. Tetsuo Kogawa, "Parts List of the Most Simplest Transmitter," *Poylmorphous Space*, http://www.translocal.jp/radio/micro/parts_simplestfm/index.html (accessed November 21, 2008).

2. Tetsuo Kogawa, "A Micro Radio Manifesto," *Poylmorphous Space* (May 22, 2006), http://anarchy.translocal.jp/radio/micro/index.html (accessed December 30, 2008).

3. Washington Post, comp. "911 Calls from September 11, 2001," *Washington Post*, 31 March 2006), http://www.washingtonpost.com/wp-dyn/content/article/2006/03/31/AR2006033100978.html (accessed December 30, 2008).

4. Gilles Deleuze, "Control and Becoming" in *Negotiations* (New York: Columbia University Press, 1995), 175.

5. Canada, *Broadcasting Act*, Part 1, Interpretation (1991), http://laws.justice.

gc.ca/en/B-9.01/ (accessed January 14, 2009).

6. Tom Schreiner quoted in Jesse Walker, in *Rebels on the Air: An Alternative History of Radio In America* (New York, NY: UP, 2001), 223.

7. Letter from Diane Rhéaume, Director General, Broadcast Analysis, CRTC to Sylvain Couture, Les émetteurs Décade (November 25, 1994), http://www.decade.ca/CRTC.html (accessed January 14, 2009). Les émetteurs Décade (Decade Transmitters) is a Canadian manufacturer and supplier of LPFM transmitters.

8. Arthur Kroker quoted in Nancy Shaw, "Cultural Democracy and Institutionalized Difference," in *Canadas*, ed. Jordan Zinovich (Brooklyn, NY: Semiotexte/Marginal, 1994), 241.

9. Charles, Fairchild, "The Canadian Alternative: A Brief History of Unlicensed and Low Power Radio," in *Seizing the Airwaves: A Free Radio Handbook*, ed. Ron Sakolsky & Stephen Dunifer (San Francisco, CA: AK Press), 1998.

10. Joseph Goebbels, "The Radio as the Eighth Great Power," in *German Propaganda Archive* (1999), http://www.calvin.edu/academic/cas/gpa/goeb56.htm (accessed December 27, 2008).

11. Jody Berland, "Contradicting Media: Toward a Political Phenomenology of Listening," in *Radiotext(e)* (New York: Semiotext(e), 1993), 209.

12. Félix Guattari, "Popular Free Radio," in *Radiotext(e)* (New York: Semiotext(e), 1993), 85.

13. Shane Breaker, "A Voice: North American Aboriginal Radio," *RIXC READER*, http://rixc.lv/reader/txt/txt.php?id=232&l=en (accessed January 15, 2009).

14. "Exemption Order Respecting Low-Power Radio: Ultra Low-Power Announcement," Public Notice CRTC 1993-46 (April 30,1993). This order exempts certain micro-radio "undertakings" — specifically, broadcasts by business interests and the authorities — from certain prohibitive clauses in the Broadcasting Act. Interestingly, the "talking signs" image is exemplary of the type of authoritarian, one-sided "apparatus of distribution" Brecht warned against. Micro-powered stations may not "let the listener speak" unless they are willing to vertically-integrate with the state under "Community Radio" provisions.

15. Karl Marx, *Capital*, (Letchworth, UK: The Temple Press, 1951), 58.

16. Canada, Industry Canada, Radiocommunications and Broadcasting Regulatory Branch, RIC-40: Frequently Asked Questions on Low Power FM Broadcasting (June 3, 2008), http://www.ic.gc.ca/eic/site/smt-gst.nsf/eng/sf02087.html (accessed January 14, 2008).

17. Marian van der Zon, "Broadcasting on Our Own Terms: Temporary Autonomous Radio," in *Autonomous Media: Activating Resistance & Dissent*, eds. Andrea Langlois and Frédéric Dubois. (Montréal, Québec: Cumulus Press, 2005), 38, http://www.cumuluspress.com/autonomous media (accessed May 12, 2009).

18. Félix Guattari quoted in Michael Goddard, "Félix and Alice in Wonderland: The Encounter between Guattari and Berardi and the Post-Media Era,"

Generation Online, http://www.generation-online.org/p/fpbif01.htm (accessed January 14, 2009).

19. Mbanna Kantako quoted in Jesse Walker, *Rebels on the Air: An Alternative History of Radio In America* (New York, NY: UP, 2001), 209.

20. Mutebutton Slim, "Kinda the Blues Every Monday 7:00–10:00 pm," (weblog entry) Tree Frog Radio 89.1 FM, (August 10, 2007), http://treefrogradio.blogspot.com/2007/08/so-tell-me-what-your-show-is-about.html (accessed January 30, 2009, currently unavailable).

21. Mary Vipond, *Listening In: The First Decade of Canadian Broadcasting, 1922-1932* (Montreal, Quebec: McGill UP, 1992), 135-136. Thanks to Marian van der Zon for this gem.

22. Chapter 7 in this volume.

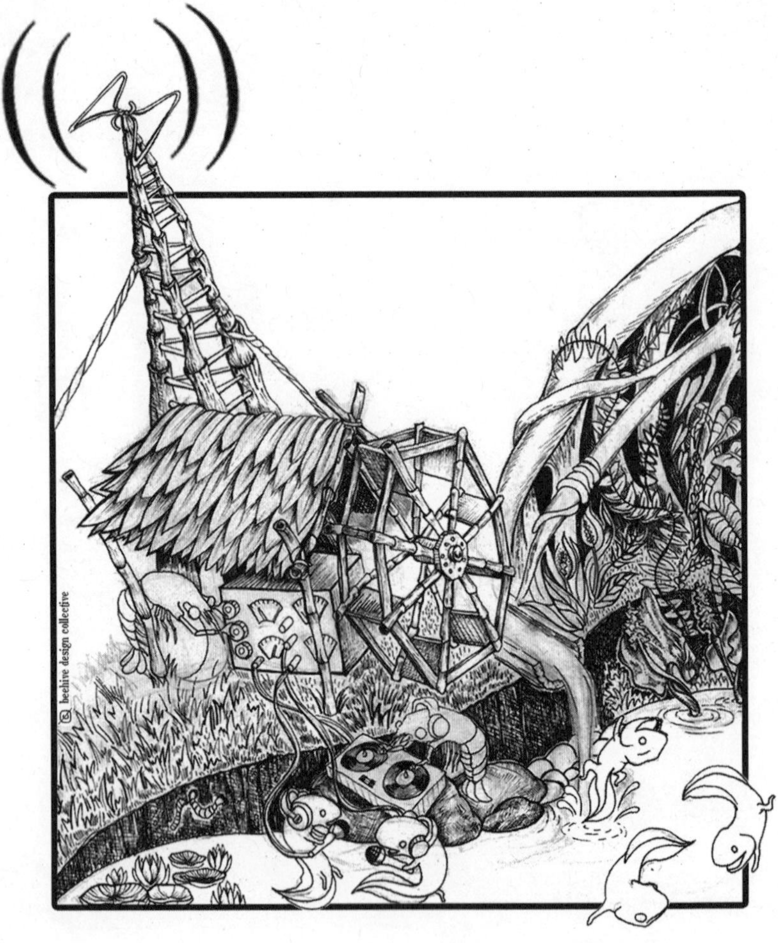
© beehive design collective

Rip-Roaring Radical Radio Resources

PIRATE RADIO DIY GUIDES

- Anonymous. *Radio Is My Bomb: A DIY Manual for Pirates*. London: Hooligan Press, 1987, or available from http://www.roguecom.com/rogueradio/radioismybomb.html.
- Disruption. *The Goal is as Real as We Make It!: Organize Radio Revolt* (an undated tech zine) Free Radio Twin Cities Collective: Minnesota, USA.
- Dunifer, Stephen. "Micropower Broadcasting: A Technical Primer." In *Seizing the Airwaves: A Free Radio Handbook*. Eds. Ron Sakolsky and Stephen Dunifer. San Franscisco: AK Press, 1998, or available at www.freeradio.org.
- Enrile, TJ. *A Popular Guide to Building a Community FM Broadcast Station*. Free Radio Berkeley, 2005 (in English or Spanish) or available from www.freeradio.org (all text and illustrations by T. J. Enrile).
- Kozinuk, Rob. "How to Build a One-Watt FM Transmitter Based on a Workshop by Tetsuo Kogawa." In *Radio Rethink: Art, Sound and Transmissions*. Eds. Daina Agaitus and Dan Lander. Banff, BC: Banff Centre for the Arts, 1994.
- Leplae, Xav. "Mutiny on the Airwaves or, How to be a Radio Pirate." An undated article published by Paper Tiger Television, New York City.
- radio4all.net. Home of the A-Infos radio project that exists "to be an alternative to the corporate and government media which do not serve struggles for liberty, justice and peace, nor enable the free expression of creativity."
- Teflon, Zeke. *The Complete Manual of Pirate Radio*. Tucson, Arizona: See Sharp Press, 1993.
- Tetsuo Kogawa. *Polymorphous Space*. http://anarchy.translocal.jp/.

- Yoder, Andrew. *Pirate Radio Stations: Tuning in to Underground Broadcasts.* Summit, PA: TAB Books, 1990.

FILMS

- *Un Poquito de Tanta Verdad* (A Little Bit of So Much Truth), produced and directed by Jill Irene Freidberg, Corrugated Films, 2008.
- *Pirate Radio USA*, produced and directed by Mary Jones and Jeff Pearson, Deface the Nation Films, 2006.
- *Rebel Radio*, produced by NEXT TV, City TV, Toronto, March 2002.
- *Free Radio*, produced and directed by Kevin Keyser (available from freeradio.org), 2000.
- *Born in Flames*, produced and directed by Lizzie Borden, First Run Features, 1983.

TRANSMITTERS

- Freeradio.org. http://www.freeradio.org/, Complete FM broadcast transmitter packages for purchase.
- Les émetteurs Décade (Decade Transmitters). http://www.decade.ca, Les émetteurs Décade is a Canadian manufacturer and supplier of LPFM transmitters.
- Quality Kits website. http://www.qkits.com/, A website where you can purchase a variety of low power FM kits and equipment.
- USB transmitter. http://www.canakit.com/, Information and equipment for USB FM transmitters.
- Veronika: AAREFF Transmission Systems. http://www.veronica.co.uk, Veronika offers 1-watt, 12-watt, 30-watt and 100-watt transmitters and antennas.

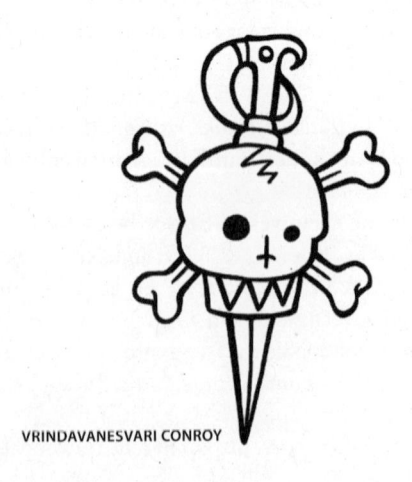

VRINDAVANESVARI CONROY

Bibliography

Adorno, Theodor. *Le caractère fétiche dans la musique*. Paris: Éditions Allia, 2007.
Anonymous. *Radio is My Bomb*. London: Hooligan Press, 1987.
Anonymous. *Radio Art: The End of the Graven Image*. Amsterdam: Gallery 'A', 1982.
Arns, Inke. "The Realization of Radio's Unrealized Potential: Media-Archaeological Focuses in Current Artistic Practice." In *Re-Inventing Radio: Aspects of Radio as Art*, ed. Heidi Grundmann et al. Frankfurt am Main, Germany: Revolver, 2008.
Augaitus, Daina and Dan Lander, eds. *Radio Rethink: Art, Sound and Transmission*. Banff: Banff Centre for the Arts, 1994.
Augoyard, Jean-Francois and Henry Torgue. *Sonic Experience*. Montreal: McGill-Queens University Press, 2005.
Bagdikian, Ben H. *The Media Monopoly*. Boston: Beacon Press, 1997.
Barlow, William. *Voice Over: The Making of Black Radio*. Philadelphia: Temple University Press, 1999.
Barthes, Roland. *A Lover's Discourse: Fragments*. Trans. Richard Howard. New York: Hill and Wang, 1993.
Becker, Howard S. *Art Worlds*. Berkeley: University of California Press, 2008.
Becker, Howard S. *Outsiders*. Paris: Minuit, 1985.
Berardi, Franco (Bifo). *Félix Guattari: Thought, Friendship and Visionary Geography*. New York: Macmillan Publishers, 2008.
Berardi, Franco M., Jacquenot and G. Vitali. "Italian Media Activism in the 1970s." In *Ethereal Shadows: Communications and Power in Contemporary Italy*. New York: Autonomedia, 2009.

Berland, Jody. "Contradicting Media: Toward a Political Phenomenology of Listening." In *Radiotext(e)*. New York: Columbia University Press, 1993.

Bey, Hakim. *The Temporary Autonomous Zone, Ontological Anarchy, Poetic Terrorism*. Brooklyn, NY: Autonomedia, 1991. http://www.to.or.at/hakimbey/taz/taz.htm (accessed January 22, 2009).

Breaker, Shane. "A Voice, North American Aboriginal Radio." *RIXC Reader*. http://www.rixc.lv/reader (accessed June 4, 2009).

Brecht, Bertolt. "The Radio as an Apparatus of Communication." (1932) In *Radiotext(e)*, ed. Neil Strauss. New York: Columbia University Press, 1993.

Brown, Carla. "Pirate Radio: A Voice for the Disenfranchised." *Peace and Environment News*. July-August, 1996. http://www.perca.ca/PEN/1996-07-08/s-brown.htm (accessed September 4, 2009).

Burroughs, William and Daniel Odier. *The Job: Interviews with William S. Burroughs*. New York: Penguin Books, 1989.

Canadian Radio-television and Telecommunications Commission. "Broadcasting sector." http://www.crtc.gc.ca/eng/bctg-radio.htm (accessed August 22, 2009).

Canadian Radio-television and Telecommunications Commission. "Public Notice 2000-10." http://www.crtc.gc.ca/eng/archive/2000/PB2000-10.htm (accessed August 22, 2009).

Carpenter, Sue. *40 Watts From Nowhere: A Journey Into Pirate Radio*. New York: Scribner, 2004.

Carter, Angela. "Preface to Come Unto These Yellow Sands." In *The Curious Room: Plays, Film Scripts and an Opera*. London: Vintage, 1997.

Chapman, Bruce. *Selling the Sixties: The Pirates and Pop Music Radio*. New York: Routledge, 1992.

Chapman, Owen. "Radio Activity: Articulating the Theremin, Ondes Martenot and Hammond Organ." *Wi: Journal of Mobile Media*. http://wi.hexagram.ca (accessed April, 2009).

Chassagne, Boris. "Quand l'art-radio bousille la colle." *Musicworks*, no. 53, 1993.

Chion, Michel. "Audio-Vision Nathan-Université." Série *Cinéma et Image*. Paris, 2005.

Collin, Matthew. *This is Serbia Calling: Rock 'N 'Roll Radio and Belgrade's Underground Resistance*. London: Serpents Tail, 2001.

Collis, Stephen. *Blackberries*. Toronto: Book Thug, 2005/06.

Coopman, Ted. "Defining Public Interest in the Micro Radio Debate: Canadian vs US Policies." 1999. http://www.roguecom.com/rogueradio/defining.html (accessed September 4, 2009).

Couldry, Nick. "Being Elsewhere: The Politics and Methods of Researching Symbolic Exclusion." In *The Politics of Place*, ed. Cresswell, T. and Verstraete, G. Amsterdam: University of Amsterdam/Rodopi Press, 2003.

D'arcy, Margaretta. *Galway's Pirate Women, a Global Trawl*. Galway: Women's Pirate Press, 1996.

Deleuze, Gilles. "Control and Becoming." In *Negotiations*. New York: Columbia University Press, 1995.

Deleuze, Gilles. *Mille plateaux : capitalisme et schizophrénie*. Paris: Minuit,1972.
Department of Justice Canada. "Radiocommunication Act." 15 June 2009. http://laws.justice.gc.ca/en/ShowFullDoc/cs/R-2//20090707/en (accessed August 22, 2009).
Depuis-Déri, Francis. *Les black-blocs: La liberté et l'égalité se manifestent*. Montréal, Québec: Lux 2003.
Doctorow, Cory. *Little Brother*. New York: Macmillan/Tor-Forge Books, 2008.
Duffy, Dennis. *Imagine Please: Early Radio Broadcasting in British Columbia*. Victoria: Provincial Archives of British Columbia, 1983.
Dunbar-Hester, Christina. "Geeks, Meta-Geeks, and Gender Trouble: Activism, Identity and Low-power FM Radio." *Social Studies of Science*, 38-2 (April 2008).
Edmonson, Richard. *Rising Up: Class Warfare in America From the Streets to the Airwaves*. San Francisco: Librad Press, 2000.
Enrile, T.J. *A Popular Guide to Building a Community FM Broadcast Station*. Berkeley California: Free Radio Berkeley, 2005.
Fabozzi, Paul F. *Artists Critics Context*. New Jersey: Prentice Hall, 2002.
Fairchild, Charles. "Below the Hamlin Line: CKRZ and Aboriginal Cultural Survival." *Canadian Journal of Communication*, vol. 23, No. 2 (1998). http://www.cjc-online.ca/index.php/journal/article/view/1031/937 (accessed September 4, 2009).
Fairchild, Charles. "The Canadian Alternative: A Brief History of Unlicenced and Low Power Radio." In *Seizing The Airwaves: A Free Radio Handbook*, ed. Ron Sakolsky and Stephen Dunifer. San Francisco: AK Press, 1998.
Felson, Marcus. *Crime and Nature*. London: Sage, 2006.
Flam, Jack. *Robert Smithson: The Collected Writings*. Berkeley: University of California Press, 1996.
Fowler, Gene and Bill Crawford. *Border Radio: Quarks, Yodelers, Pitchmen, Psychics, and Other Amazing Broadcasters of the American Airwaves*. Austin: University of Texas Press, 2002.
Freud, Sigmund. *L'inquiétante étrangeté et d'autres essais*. Paris: Gallimard, 1988.
Friz, Anna. "Radio As Instrument." *Wi: Journal of Mobile Media*. http://wi.hexagram.ca (accessed April, 2009).
Girard, Bruce, ed. *A Passion for Radio*. Montréal: Black Rose Books, 1992.
Goddard, Michael. "Félix and Alice in Wonderland: The Encounter between Guattari and Berardi and the Post-Media Era." *Generation Online*, http://www.generation_online.org/p/fpbifo1.htm (accessed August 22, 2009).
Goebbels, Joseph. "The Radio as the Eighth Great Power." In *German Propaganda Archive* (1999). http://www.calvin.edu/academic/cas/gpa/goeb56.htm (accessed December 27, 2008).
Goldberg, David. "The Scratch is Hip Hop: Appropriating the Phonographic Medium." In *Appropriating Technology: Vernacular Science and Social Power*, ed. Ron Eglash. Minneapolis: University of Minnesota Press, 2004.
Graeber, David. *Direct Action: An Ethnography*. Oakland, CA: AKPress, 2009.

Graeber, David. "The New Anarchists." *The New Left Review.* 13, 2002.
Grundmann, Heidi, Elisabeth Zimmermann, Reinhard Braun, Dieter Daniels, Andreas Hirsch, Anne Thurmann-Jajes, eds. *Re-Inventing Radio: Aspects of Radio as Art.* Frankfurt: Revolver, 2008.
Guattari, Félix. "Popular Free Radio." In *Radiotext(e).* New York: Semiotext(e), 1993.
Guattari, Félix. *Soft Subversions,* ed. Sylvere Lotringer. New York: Semiotext(e), 1996.
Harney, Stefano. "Governance and the Undercommons." 2008. http://info.interactivist.net/node/10926 (accessed August 22, 2009).
Hilmes, Michele. *Radio Voices: American Broadcasting, 1922-1952.* Minneapolis: University of Minnesota Press, 1997.
Hind, John and Stephen Mosco. *Rebel Radio: The Full Story of British Pirate Radio.* London: Pluto Press, 1985.
Industry Canada, Spectrum Management and Telecommunications, "BPR (broadcasting procedures and rules) -1 General Rules." Issue 5, January 2009. http://www.ic.gc.ca/eic/site/smt-gst.nsf/eng/sf01326.html (accessed August 22, 2009).
Jensen, Erik Granly and Brandon La Belle, eds. *Radio Territories.* Los Angeles: Errant Bodies Press, 2007.
Jeschke, Rebecca. "EFF Tackles Bogus Podcasting Patent - And We Need Your Help," Electronic Frontier Foundation, November 19, 2009. http://www.eff.org/deeplinks/2009/11/eff_tackles_boguspodcasting_patent_and_we_need_yo (accessed December 12, 2009).
Joyce, Zita. "Electromagnetic myths, ether vibrations in the space between the worlds." In *Spectropia: Illuminating Investigations in the Electromagnetic Spectrum,* ed. Daina Silina et al. Riga, Latvia: RIXC Centre for New Media Culture, 2008.
Kahn, Doug and Gregory Whitehead. *Wireless Imagination: Sound, Radio and the Avant Garde.* Cambridge, MA: MIT Press, 1992.
Keith, Michael. *Signals in the Air: Native Broadcasting in America.* Westport, Connecticut: Praeger Press, 1995.
Kogawa, Tetsuo. "A Micro Radio Manifesto." *Polymorphous Space.* 2006. http://anarchy.translocal.jp/radio/micro/ (accessed August 22, 2009).
Kogawa, Tetsuo. "How to build a micro transmitter." *Polymorphous Space.* 2006. http://anarchy.translocal.jp/radio/micro/howtotx.html (accessed March 15, 2009).
Kogawa, Tetsuo. "Mini-FM: Performing Microscopic Distance." In *At A Distance: Precursors to Art and Activism on the Internet,* ed. Annmarie Chandler and Norie Neumark. Cambridge, Mass: The MIT Press, 2005.
Kogawa, Tetsuo. "Radio in the Chiasme." *Re-inventing Radio: Aspects of Radio as Art,* Heidi Grundmann et al, eds. Frankfurt Am Main: Revolver, 2008.
Kogawa, Tetsuo. "Parts List of the Most Simplest Transmitter." *Poylmorphous Space.* 2006. http://www.translocal.jp/radio/micro/parts_simplestfm/index.html (accessed November 21, 2008).

Kogawa, Tetsuo. "Toward Polymorphous Radio." In *Radio Rethink: Art, Sound and Transmission*, ed. Daina Augaitis and Dan Lander. Banff: Walter Philips Gallery, 1994.
Kogawa, Tetsuo. "What is Radio Party?" *Polymorphous Space*. 2006. http://anarchy.translocal.jp/radio/radioparty/index.html (accessed February 4, 2009).
Kuhn, Gabriel. "Life Under the Death's Head: Anarchism and Piracy." In *Women Pirates and the Politics of the Jolly Roger*, eds. Ulrike Klausmann, Marion Meinzerin, and Gabriel Kuhn. Montréal: Black Rose Books, 1997.
LaBelle, Brandon. *Background Noise*. New York: Continuum, 2006.
LaBelle, Brandon. "Transmission Culture." In *Re-Inventing Radio: Aspects of Radio as Art*, ed. Heidi Grundmann et al. Frankfurt am Main, Germany: Revolver, 2008.
Land, Jeff. *Active Radio: Pacifica's Brash Experiment*. Minneapolis: University of Minneapolis Press, 1999.
Lander, Dan. *Selected Survey of Radio Art in Canada, 1967-1992*. Banff, Alberta: Walter Phillips Gallery, 1994.
Langlois, Andrea and Frédéric Dubois, eds. *Autonomous Media: Activating Resistance and Dissent*. Montréal: Cumulus Press, 2005. http://www.cumuluspress.com/autonomousmedia.html (accessed September 11, 2009).
Langlois, Andrea M. *Mediating Transgressions: The Global Justice Movement and Canadian News Media*. Unpublished Master's thesis: Concordia University, 2004.
Lasar, Matthew. *Pacifica Radio: The Rise of an Alternative Network*. Philadelphia: Temple University Press, 1999.
Létourneau, Éric André. "L'empire des ondes." *Parallélogramme*. 16, no.4, Toronto: Éditions ANPAC\RACA, 1991.
MacLennan, Anne F. "Cultural Imperialism of the North? The Expansion of CBC's Northern Service." *The Radio Journal*. Forthcoming.
Marx, Karl. *Capital*. Letchworth, UK: The Temple Press, 1951.
Medosch, Armin et al, eds. *Waves: Electromagnetic Waves as Material and Medium for Arts*. Acoustic Space Lab #6. Riga, Latvia: RIXC Centre for New Media Culture, 2006.
Mercure, Pierre. "Commentaires." In *Musique du Kébèk*, ed. Raoul Duguay. Montréal: Éditions du jour, 1971.
Milam, Lorenzo Wilson. *Sex and Broadcasting: A Handbook on Starting a Radio Station for the Community*. San Francisco: Dildo Press, 1975.
Milam, Lorenzo Wilson. *The Radio Papers*. San Diego: MHO and MHO Works, 1986.
Milatus, Joe. "Radiophonic Ontologies and the Avante Garde." In *Experimental Sound and Radio*, ed. Allan S. Weiss. TDR Books: New York, 2001.
Mitchell, Caroline, ed. *Women & Radio: Airing Differences*. London: Routledge, 2000.
Morris, Adalaide ed. *Sound States: Innovative Poetics and Acoustical Technologies*. Chapel Hill and London: University of North Carolina Press, 1997.
Mostoller, Charles. "Oaxaca's Media Wars." *Znet*. http://www.zmag.org/znet/

viewArticle/17490 (accessed May 10, 2009).

Mudede, Charles. "The Turntable." In *Life In The Wires: A CTheory Reader*. Canada: New World perspectives / CTheory Books, 2004.

"The Future of Radio Issue." *Musicworks* No. 106, Spring 2010, Toronto.

Nopper, Sheila. "People Have No Idea How Powerful They Could Be: An Interview with Carol Denney (Free Radio Berkeley)." In *Seizing The Airwaves: A Free Radio Handbook*, ed. Ron Sakolsky and Stephen Dunifer. San Francisco, California: AK Press, 1998.

Nopper, Sheila. "We Have to Make Sure That The Voiceless Have A Voice: An Interview with Kiilu Nyasha (San Francisco Liberation Radio)." In *Seizing The Airwaves: A Free Radio Handbook*, ed. Ron Sakolsky and Stephen Dunifer. San Francisco, California: AK Press, 1998.

Olivelucy and Salmonella. "Riding Radio to Choke the IMF." In *That's Revolting: Queer Strategies for Resisting Assimilation*, ed. by Mattilda (aka Matt Bernstein Sycamore). Brooklyn, New York, Soft Skull Press, 2004.

Page, Tim. *The Glenn Gould Reader*. New York: Knopf, 1984.

Pelletier, Sonia. "Ondes fluides et points de force." *Inter Inter*, no.55, Quebec City: Éditions Interventions, 1993.

Peters, John Durham. *Speaking into the Air: A History of the Idea of Communication*. Chicago: University of Chicago Press, 2000.

Raboy, Marc. *Missed Opportunities: The Story of Canada's Broadcasting Policy*. Montréal and Kingston: McGill-Queens University Press, 1990.

Rasmussen, Mikkel Bolt. "Promises in the Air: Radio Alice and Italian Autonomia." In *Radio Territories*, ed. Erik Granly Jensen and Brandon LaBelle. Los Angeles: Errant Bodies Press, 2007.

Richard, Alain Martin. "énoncés généraux-Matériau: manœuvre." *Inter, #47*. Québec: Éditions intervention, 1990.

Roos, Kristen. "The Parking Lot Broadcast." In *Radio Territories*, ed. Erik Granly Jensen and Brandon LaBelle. Los Angeles: Errant Bodies Press, 2007.

Ruggiero, Greg. *Microradio and Democracy: (Low) Power to the People*. New York: Seven Stories Press, 1999.

Sakolsky, Ron. "Insurrectionary Radio." In *Swift Winds*. Portland, Oregon: Eberhardt Press, 2009.

Sakolsky, Ron. "The Myth of Government-Sponsored Revolution: A Cautionary Tale." In *Creating Anarchy*. Liberty, Tennessee: Fifth Estate Books, 2005.

Sakolsky, Ron and Stephen Dunifer, eds. *Seizing the Airwaves: A Free Radio Handbook*. San Francisco: AK Press, 1998.

Sanouillet, Michel. *Salt Seller: The Writings of Marcel Duchamp*. New York: Oxford University Press, 1973.

Sey, James. "Sounds Like...: the Cult of the Imaginary Wavelength." In *Radio Territories*, ed. Erik Granly Jensen and Brandon LaBelle. Los Angeles and Copenhagen: Errant Bodies Press, 2007.

Shukaitis, Stevphen. "Dancing Amidst The Flames: Imagination and Self-Organization in a Minor Key." In *Subverting The Present, Imagining the Future: Insurrections, Movement, Commons*, ed. Werner Bonefield. New York:

Autonomedia Press, 2008.
Shukaitis, Stevphen. *Imaginal Machines: Autonomy and Self-Organization in the Revolution of Everyday Life*. Brooklyn, NY: Autonomedia Press, 2009.
Shaw, Nancy. "Cultural Democracy and Institutionalized Difference." In *Canadas*, ed. Jordan Zinovich. Brooklyn, NY: Semiotexte/Marginal, 1994.
Silberman, Marc. *Brecht on Film and Radio*. London: Methuen Press, 2000.
Squier, Susan Merrill. *Communities of the Air: Radio Century*, Radio Culture. Durham, NC: Duke University Press, 2003.
Stevensen, John Harris, Tristis Ward and Melissa Kaestner. "Some NCRA/ANREC History." http://www.ncra.ca/business/History.cfm (accessed September 4, 2009).
Strauss, Neil and Dave Mandl, eds. *Radiotext(e)*. New York: Semiotext(e), 1993.
Soley, Lawrence. *Free Radio: Electronic Civil Disobedience*. Boulder: Westview Press, 1999.
Stanley, Jo, ed. *Bold in Her Breeches: Women Pirates Across the Ages* (San Francisco, California: Pandora, 1995.
Tyson, Timothy. *Radio Free Dixie: Robert F. Williams and the Roots of Black Power*. Chapel Hill: University of North Carolina Press, 1999.
van der Zon, Marian. "Broadcasting On Our Own Terms: Temporary Autonomous Radio." In *Autonomous Media: Activating Resistance & Dissent*, eds. Andrea Langlois and Frédéric Dubois. Montréal, Québec: Cumulus Press, 2005. http://www.cumuluspress.com/autonomous media (accessed May 12, 2009).
Vigil, José Ignacio López. *Rebel Radio: The Story of El Salvador's Radio Venceremos*. Wilimantic, Conn: Curbstone Press, 1994.
Vipond, Mary. *Listening In: The First Decade of Canadian Broadcasting, 1922-1932*. Montréal, Québec: McGill UP, 1992.
Walker, Jesse. *Rebels on the Air: An Alternative History of Radio in America*. NYC: New York University, 2001.
Ward, Colin. *Anarchy in Action*. London: Freedom Press, 1973/82.
Weiner, Allan H. *Access to the Airwaves: My Fight for Free Radio*. Port Townsend, WA: Loopanics Unlimited, 1997.
Weiss, Allen. *Phantasmic Radio*. Durham, NC: Duke University Press, 1995.
Weiss, Allen, ed. *Experimental Sound and Radio*. Cambridge, MA: MIT Press, 2001.
Yoder, Andrew. *Pirate Radio Stations: Tuning in to Underground Broadcasts*. Summit, PA: TAB Books, 1990.
Yoder, Andrew. *Pirate Radio*. Solana Beach, CA: Hightext Publications, Inc, 1996.
Yoder, Andrew and Earl Gray. *Pirate Radio Operations*. Port Townsend, WA: Loompanics Unlimited, 1997.

The Contributors

STEPHEN DUNIFER is known in the United States as the "Johnny Appleseed" of the micropower broadcasting movement. He is the founder of the legendary Free Radio Berkeley, and co-editor (with Ron Sakolsky) of the book, *Seizing The Airwaves: A Free Radio Handbook* (AK Press, 1998). He is the organizer of International Radio Action Training in Education (IRATE) and Project TUPA (Transmitters Uniting the Peoples of the Americas).

ROGER FARR is the author of a book of poetry, *Surplus* (LineBooks, 2006), a contributor to the co-research project *N 49 19. 47 – W 123 8.11* (Recomposition, 2008), and the editor of *PARSER: New Poetry and Poetics*. Recent writing appears or is forthcoming in *Anarchist Studies, Fifth Estate, Politics is Not a Banana, The Post-Anarchism Reader, Rad Dad, Social Anarchism*, and *XCP: Cross Cultural Poetics*. He teaches in the Creative Writing and Culture and Technology Programs at Capilano University in Vancouver, BC, but lives on an island at the end of the dial, somewhere in the Salish Sea.

ANNA FRIZ is a sound and radio artist, and a critical media studies scholar. Since 1998 she has predominantly created self-reflexive radio art/works for broadcast, installation or performance, where radio is the source, subject, and medium of the work. She has extensively performed and exhibited installation works at festivals and venues across Canada, the US, in Mexico, and across Europe. Her radio art/works have been commissioned by national public radio in Canada, Austria, Germany, Denmark, and Mexico, and heard on independent airwaves in more than 20 countries. Anna Friz is a free103point9.org transmission artist, and is completing her doctoral degree in Communication and Culture at York University, Toronto. http://nicelittlestatic.com

STEPHEN KELLY is an artist who works with sound, installation, electronic/computational art, and low power FM radio. He has exhibited and participated in residency programs both nationally and internationally. Interested in the intersections between audio art and music, Stephen builds unique musical instruments and approaches sound recording as a creative process. His most recent musical project, in collaboration with Eleanor King, is entitled *The Just Barelys*. Stephen has a BFA from the Nova Scotia College of Art & Design and is currently studying computer science at Dalhousie University.

KATHY KENNEDY is a sound artist with formal training in visual art as well as classical singing. Her practice generally involves the voice and issues of interface with technology, telephony or radio transmission. She is a founder of Studio XX in Montréal, and director of many ongoing choral and community projects and site-specific installations.

GRETCHEN KING has been cultivating spaces for Indymedia radio mobilizations since the WTO dared to meet in Seattle in 1999. She has been an active participant in radio revolution as a means of connecting mobilizations worldwide through the FM dial and over the Internet. Since 2001, Gretchen has served as Community News and Production Coordinator at CKUT Radio (90.3 FM) in Montreal. In addition to amplifying local resistance movements with Radio Taktic and Sonique Resistance, she has also coordinated Canada's annual Homelessness Marathon for the last eight years and co-founded *GroundWire*, a national grassroots news magazine.

ELEANOR KING is an interdisciplinary installation and performance artist who fuses found materials in a playful way to critique social behaviors, investigating cultures of consumerism and tourism. She has exhibited nationally and internationally, and has participated in residency programs in Canada and the US. Eleanor currently teaches in the Media Arts department at NSCAD University, holds the position of Exhibitions Coordinator at Anna Leonowens Gallery, and is also a member of indie-rock bands The Just Barelys and The Got To Get Got.

BOBBI KOZINUK is a Vancouver-based media artist, technician and curator. Former Media Director at the Western Front, Bobbi has worked on a board level with the Independent Media Arts Alliance (Montréal), Co-op Radio, Grunt Gallery and Video In (Vancouver) and has traveled extensively producing workshops on low-powered FM transmission across Canada at artist-run centres. Bobbi is published in *Radio Rethink* (produced by the Banff Centre for the Arts) and *Echo Locations* (Audio Art CD produced by Co-op Radio). Currently the electronics studio technician at the Emily Carr University of Art and Design, Bobbi has exhibited media installation works in both national and international contexts including Diffractions, Galleria di Nuova Icona, Venice, Italy, and Folly Gallery, Lancaster, UK. Bobbi teaches electronics for artists at the University of British Columbia.

ANDRÉ ÉRIC LÉTOURNEAU is an artist, author and active participant in the worlds of interdisciplinary art, media arts and radio art. Since the 1980s, his maneuvers, installations, concerts and radiophonic creations have been presented at more than 50 events, festivals and museums around the world. His writings cover art, culture, politics and criminology, and have been published in *Éditions Interventions, Esse, ANPAC\RACA, Artexte, Art Métropole*, on the Radio-Canada website, and in editions of the l'Académie Non Grata (en Estonie) and the Université de Montréal press. He is also active as a programming and organizational planning consultant for various organizations including Articule, Dare–Dare, the RAIQ, the Biennale de Paris, the Canadian Council for the Arts, and the Conseil des Arts de Montréal.

ANNE MacLENNAN is an assistant professor in the Department of Communication Studies at York University and the York–Ryerson Joint Graduate Program in Communications and Culture. She is a media historian whose research focuses primarily on Canadian radio programming and audiences during the 1930s. The remainder of her teaching and research intersects with a variety of interdisciplinary media studies of radio, television, advertising, women, labour and social welfare.

NESKIE MANUEL got his first taste of radio at CFBX in Kamloops, BC in Secwepemcu'lucw, his homeland. There he produced a show that highlighted music whether it be indigenous or not, but more importantly which spoke to the events of the day. When his father started up an unlicensed radio station on the Neskonlith Reserve, Neskie helped with the operations, from making coffee for volunteers to fixing the transmitter.

CHRISTOF MIGONE is a multidisciplinary artist and writer who worked extensively in radio from 1984 to 1994, primarily at CKCU–FM (Ottawa), CKUT–FM (Montréal), and Radio Zones (Ferney-Voltaire, France). He co-edited the book and CD *Writing Aloud: The Sonics of Language* (Los Angeles: Errant Bodies Press, 2001) and his writings have been published in *Aural Cultures, S:ON, Experimental Sound & Radio, Musicworks, Radio Rethink, Semiotext(e), Angelaki, Esse, Inter*, etc. He obtained an MFA from NSCAD in 1996 and a PhD from the Department of Performance Studies at the Tisch School of the Arts of New York University in 2007. He currently lives in Toronto and is a lecturer at the University of Toronto Mississauga and the Director/Curator of the Blackwood Gallery.

CHARLES MOSTOLLER is a photojournalist and writer who focuses on issues of social justice, especially the struggle for indigenous and immigrant rights. He is a freelance photographer for the daily newspaper *Metro* in New York, has worked for both English and Spanish language community radio stations in Montréal, Québec and has published articles with alternative media outlets such as *AlterNet, The Dominion, CounterPunch*, and *ZNet*. Charles lived and reported from Mexico between 2006 and 2008, where he wrote in-depth reports on a number of topics, including the current wave of community radio stations being created in indigenous communities in the southern state of Oaxaca. Back in Canada, he joined an indigenous solidarity group called Barriere Lake Solidarity, which works to help the Algonquin of Barriere Lake in their struggle to obtain rights over their traditional territory. Charles helped the Algonquin of Barriere Lake establish a Low Power FM radio station in the community, which broadcasts in the Algonquin language. His website is www.charlesmostoller.com/blog.html

SHEILA NOPPER has been a community radio DJ, interviewer and documentary producer at CIUT in Toronto, and a correspondent for CBC's *Global Village*. While living in Illinois, she founded the Media Activist Coalition, and produced *Grrrrl Radio*, a live broadcast on the educational station WQNA featuring girls aged 8–14. She has an MA in Media and Cultural Activism, and has written for such publications as *Herizons*, *Fifth Estate*, *Illinois Times* and *The Beat*, as well as being featured in the anthology, *Seizing the Airwaves: A Free Radio Handbook*. She currently broadcasts the show *freedom soundz* on a pirate radio station she co-founded five years ago in rural BC.

KRISTEN ROOS is a sound artist currently based in Vancouver. His work has been exhibited/performed in artist-run centres and festivals nationally, as well as being featured in the Errant Bodies publication *Radio Territories*. Kristen completed a BFA at Concordia University and an MFA at the University of Victoria. His website is www.kristenroos.com.

The editors

ANDREA LANGLOIS has been involved in autonomous media for almost ten years. Andrea was an active member of the Indymedia movement (CMAQ.net) for several years, and her voice has sailed the airwaves on many stations, from Montreal's CKUT, to Nelson's CJLY, and has been wild and free during many pirate radio broadcasts. *Autonomous Media: Activating Resistance and Dissent* (Cumulus, 2005), the book she co-edited with Frédéric Dubois, documents uses of media by social justice movements in Canada; it was published in French translation by Lux Éditeur in 2006. She works in communications in Victoria, BC, and is the editor of the quarterly magazine *BC Organic Grower*.

RON SAKOLSKY has been a radio pirate for over twenty years in both the States and Canada. With Stephen Dunifer, he co-edited the book *Seizing the Airwaves: A Free Radio Handbook* (AK Press, 1998). His writings on pirate radio have appeared in such magazines as *Fifth Estate*, *Social Anarchism*, *Cultural Democracy*, *Index on Censorship* and *Confluence*. His two most recent books are *Creating Anarchy* (Fifth Estate, 2005) and *Swift Winds* (Eberhardt Press, 2009).

MARIAN VAN DER ZON founded a pirate radio station (TAR: Temporary Autonomous Radio) in 2003, a station that is still active. She is a media activist and musician who has been involved in radio for ten years, hosting shows and contributing sound art and radio documentary pieces to CBC Radio 1, Victoria's CFUV, Montreal's CKUT, Nanaimo's CHLY, WINGS, and to numerous pirate radio stations. She teaches in the Media Studies and Women's Studies departments at Vancouver Island University. Her written work on pirate radio has been published in *Autonomous Media* (Cumulus, 2005) and its French translation, in *Social Anarchism*, and *Canadian Women in Radio* (forthcoming). She plays in the band Puzzleroot.

NEW STAR BOOKS LTD.
107 — 3477 Commercial Street | Vancouver, BC V5N 4E8 | CANADA
1574 Gulf Rd., #1517 | Point Roberts, WA 98281 | USA
www.NewStarBooks.com | info@NewStarBooks.com

Copyleft Andrea Langlois, Ron Sakolsky, & Marian van der Zon 2010. The content of this book may be reproduced without the authors' permission in part or in its entirety provided it is distributed and made available to the reader for free, without service charges or any other fee. The authors further stipulate that the editors, individual writers, and visual artists all be credited for their work.

Those wishing to reproduce any part of this work for commercial use, including course packages and other educational purposes, should contact the publisher or Access Copyright.

Publication of this work is made possible by the support of the Canada Council, the Government of Canada through the Department of Canadian Heritage Book Publishing Industry Development Program, the British Columbia Arts Council, and the Province of British Columbia through the Book Publishing Tax Credit.

Cover illustration by Maurice Spira
Printed on 100% post-consumer recycled paper
Printed and bound in Canada by Imprimerie Gauvin
First printing, May 2010

LIBRARY AND ARCHIVES CANADA CATALOGUING IN PUBLICATION

Islands of resistance : pirate radio in Canada / edited by Andrea Langlois, Ron Sakolsky and Marian van der Zon.

Includes bibliographical references.

ISBN 978-1-55420-050-4

1. Pirate radio broadcasting — Canada. 2. Pirate radio broadcasting — Social aspects — Canada. I. Langlois, Andrea II. Sakolsky, Ronald B. III. van der Zon, Marian.

HE8697.65.C3185 2010   302.23'440971   C2010-901737-4